ANTHILL ECONOMICS

Animal Ecosystems and the Human Economy

Nathanial Gronewold

Prometheus Books

Guilford, Connecticut

Ⓟ Prometheus Books

An imprint of The Rowman & Littlefield Publishing Group, Inc.
4501 Forbes Boulevard, Suite 200
Lanham, Maryland 20706
www.rowman.com

Distributed by NATIONAL BOOK NETWORK

British Library Cataloguing in Publication Information Available

Library of Congress Cataloging-in-Publication Data
Name: Gronewold, Nathanial, 1977–, author.
Title: Anthill economics : animal ecosystems and the human economy / Nathanial Gronewold.
Description: Lanham, MD : Prometheus, [2021] | Includes bibliographical references and index. | Summary: "Anthill Economics puts forth an innovative and cross-disciplinary approach, asserting that the laws of nature and physics global economy as they are to animal ecosystems. This clearly written book full of illuminating ecological analogies gives readers an informed and engaging introduction to the cutting-edge field of biophysical economics that seeks to provide a more complete understanding of the global economy"—Provided by publisher.
Identifiers: LCCN 2020035974 (print) | LCCN 2020035975 (ebook) | ISBN 9781633886520 (cloth ; alk. paper) | ISBN 9781633886537 (ebook)
Subjects: LCSH: Economics. | Animal communities.
Classification: LCC HB171 .G76 2021 (print) | LCC HB171 (ebook) | DDC 330—dc23
LC record available at https://lccn.loc.gov/2020035974
LC ebook record available at https://lccn.loc.gov/2020035975

Man has always deceived himself when he abandoned experience to follow imaginary systems.—He is the work of nature.—He exists in Nature.—He is submitted to the laws of Nature.—He cannot deliver himself from them:—cannot step beyond them even in thought. It is in vain his mind would spring forward beyond the visible world: direful and imperious necessity ever compels his return—being formed by Nature, he is circumscribed by her laws; there exists nothing beyond the great whole of which he forms a part, of which he experiences the influence.

—Paul-Henri Thiry, Baron d'Holbach,
The System of Nature (1770)

CONTENTS

INTRODUCTION

Does modern economic theory violate some very basic, fundamental laws of physics? I passed this question on to my readers several years ago. Few people bother to ask economists such a thing in polite circles, and even fewer economists feel compelled to answer them. In fact, when posed with this inquiry, most economists openly reject it, sometimes with hostility. Some in modern academia have spoken to me lately of apparently underhanded moves by the defenders of classical and neoliberal economic theory to quasi-blacklist those rare scholars in their profession who dare to muse about this question or seek out a dignified response should they come across it.

And why shouldn't most mainstream economists reject this question outright, as they apparently do? What does it mean anyway? Look at things from economists' perspective. Though its practitioners are often accused of succumbing to what critics have termed "physics envy," given their extreme fondness for complex mathematical modeling, economics is at its heart a social science. It's often characterized as a sophisticated study of supply and demand or a professional inquiry into how humans with infinite wants but finite cash and time acquire the goods and services they desire—or think they desire or are told to desire—in an open market with an abundant variety of choices. A study of trade-offs, motives, and rational decision making based on ample information. Or so the script goes.

Considering how catastrophically wrong mainstream economists have been about so many things, so many times, in so many different ways in recent memory, perhaps it's time to rewrite the script. If some radical new thinkers are to be believed, answering the question "Does economics run contrary to physics?" could be crucial to avoiding a potentially bleak future for humankind.

This question—relating the field of economics to the laws of physics and nature—was first posed to me almost a decade ago in the city of Syracuse, New York. The occasion was a small gathering of fringe scholars at the State University of New York's College of Environmental Science and Forestry to trade ideas and occasional barbs in person rather than via e-mail. The few dozen participants (about fifty or so in total) who attended this unusual, mostly invisible conference didn't drop in for the day to discuss wetland rehabilitation, endangered species habitat protection, prescribed burning, or even reforestation. The problem of global warming didn't even draw that much attention. No, the subject at hand for this room full of environmental science PhDs was neoclassical economics and one key ingredient missing from all those equations that professional economists employ, the incoherent jumble of Greek letters and other symbols that baffles the layperson. The missing factor in those equations, according to these attendees? Energy.

No, not oil and gas prices, coal demand, or local electricity rates; those topics are covered by economists all the time and in immense detail, as you may already know. The attendees of the second annual gathering of this newly formed International Society for Biophysical Economics understood this as well. No, what they had in mind was a more fundamental conceptualization of energy. Allow me to provide a very simply (borderline idiotic) illustration.

Imagine a squirrel running around in a grove in search of nuts. We've all seen this. The squirrel runs along the ground, scrambling up trees, seeking out traces, and snatching up food resources wherever and whenever she can find them. In the process she gathers dozens or even hundreds of nuts throughout the course of the day (perhaps occasionally stuffed in her cheeks) and literally squirrels them away into a little nook somewhere. Once a few are stored for safekeeping, off the little squirrel goes again to find more as quickly as possible before the winter sets in.

It's an enormous expenditure of calories, which is a measure of stored energy burned by the physical exertion of organic matter, or life. How does the squirrel recover this spent energy? Easy—she merely consumes one or several of the nuts she's been busy storing for winter this entire time. With each fresh energy boost, her frame is reinvigorated, allowing her to move again, to run even farther afield, faster perhaps, to new trees and hollows, off to newer and greater opportunities in an effort to expand her bounty as much as possible. And when she starts getting weak, she can simply consume more nuts, boosting her energy stores further for the additional work ahead.

Now, readers, let me ask you something: How long can our squirrel keep all this up?

Biologists and wildlife researchers have long known the answer, and it's rather simple on its face: the squirrel can continue expending these calories so long as her nut consumption allows her to gain more caloric energy back than the amount of energy she burned in the pursuit and acquisition of those nuts. In other words, the nuts she eats have to net her back more energy than she burned in finding and taking possession of those same nuts in the first place.

Of course, old age, injury, disease, predation, and other factors of mortality will eventually take their toll, and the squirrel's days of foraging will inevitably come to an end. But the above concept is more or less foundational to biology. Individual living creatures, and indeed entire colonies of life, can survive and thrive for only so long as they can obtain more energy from their environment than they consume in the pursuit and acquisition of this same energy from that environment. That's all a bit too wordy, so biologists have devised a clever acronym to describe the process and to save space in their academic journals: EROI, short for net energy return on investment.

This is the fundamental notion of energy and its expenditure that the women and men of forestry, ecology, biology, and hydrology had in mind at that humble gathering in Syracuse those many years ago. Hence the question they continue to challenge mainstream economics with today: If EROI is essential to all wildlife, individually and collectively, then why should humans or human societies at large be any different? Indeed, if EROI is critical to understanding ecology in general (and it is), then why would it not also apply in equal measure to the most popular form of human ecological study around today, a popular field known otherwise as economics?

That's all the world economy really is, after all, if you think about it.

Our economy is humanity's fundamental ecology. They are one and the same, perhaps best described as the economy/ecology or ecology/

economy, depending on which ordering you prefer. It describes the variety of clever methods by which humans obtain, transform, distribute, and consume resources, which require energy of all sorts: barrels of oil, tons of coal, watts of solar power, joules, British thermal units, and calories, among other measures. Eat, drink, breathe, urinate, defecate, shelter, procreate, sleep—then do it all over again the next day. This is your purpose in life. Work and trade facilitate the above. And all of it depends on energy. Everything else is just details (important details, of course, but details nonetheless).

Proponents of the new fringe field of biophysical economics understand all this in their bones and consider EROI as sacrosanct and fundamental to the global economy as it is to marine science, zoology, and entomology. Actual professional economists do not share this view, however, and indeed many of them are actually openly hostile to the very idea.

But who's ultimately right? Don't the biologists and ecologists have a rather good point? Humans are animals, after all, evolved from creatures that crawled out of the ocean many hundreds of millions of years ago. So why should we be anything special? How are we exempt from the iron-clad law of energy return on investment?

Would not a fuller understanding of and appreciation for the rule of energy return on investment improve the practice of modern economics and thus help better illuminate for all of us exactly what is happening to our world today? Heck, couldn't a little change in the weather do the economists some good anyway? Practically none of them saw the 2008–2009 global financial crisis coming. In fact few, if any, have ever accurately predicted any recession or economic depression, ever. By some twists and turns in econometric equations, recessions are rendered impossible to them, at least mathematically. But they occur nonetheless, all the time, along with a host of other modern-day economic and social phenomena that economists are now struggling to explain for us: economic stagnation, wage stagnation, rising and exacerbating inequality, plummeting birthrates, and rising hostility to so-called free trade. Some reports suggest that economists are now among the least trusted professionals in the public eye, running neck and neck with politicians and members of the press. The economists' overwhelming influence on public policy and their miserable track record in managing this immense power suggest that they've perhaps earned the public's scorn, especially for those who recall 2008 only too well. No other academic discipline in modern memory has had the capacity to destroy so many lives so swiftly as economics. Wildlife biologists certainly don't wield this immense power.

Hence the arguments of the proponents of this new field of biophysical economics and their demands for respect. The modern school of economics is flawed with regard to its thinking and understanding of energy and thus has failed, biophysical economists insist. Again, they don't mean energy as so much oil or natural gas or wind power; rather, they are referring to the fundamental importance of energy in transforming matter into food, shelter, and material possessions and the foundational role of EROI in keeping that squirrel kicking. As one scholar put it to me almost ten years ago, one possible reason why modern-day economists get society so flatly wrong so frequently is that they treat energy "the same as they would mittens or earmuffs or eggs," except without energy there can be no mittens or earmuffs or eggs (from my October 2009 interview with Nate Hagens, then editor of *The Oil Drum* blog and now a fellow at the Post Carbon Institute).

We need not single out economists for this sort of myopic tunnel vision.

Deep within the halls of academia, in the dungeon many floors beneath the ivory tower, lies a burgeoning movement of evolutionary biologists, ecologists, wildlife researchers, physicists, and even mechanical engineers pushing forward their own controversial alternative explanations for present human economic conundrums. These women and men are practitioners of fields adept at breaking down the natural world into a few fundamental laws and principles, and they're bringing these skills to bear on the confusing mess that is modern civilization. They are now challenging not only economists but also sociologists, political scientists, demographers, historians, and other so-called experts of society already given ample airtime to publicize their unimaginative, oft-repeated mainstream explanations and remedies. Taking "interdisciplinary" to a whole new level, these new rebel scientific thinkers are stepping out of their own wheelhouses of the physical and natural sciences to try their hands at dabbling in the social sciences instead, especially economics, sometimes risking professional reputations and even tenure in order to do so. And the risk is very real—modern academia can be most unforgiving to pioneers and rebels, and the ivory tower especially hates scholars who can write coherently—but we're lucky that they're taking it. Their results are at once fascinating and potentially terrifying, depending on your point of view.

This book serves as space for these ideas, a platform for their presentation. These alternative economic thinkers deserve to be heard and recognized, in my opinion, as their amazing and intuitive insights potentially hold the power to completely transform our very fundamental understanding of the human economy/ecology and how the rules of nature could make certain economic outcomes inevitable.

During this fascinating tour, you will bear witness to a set of ideas that are not all that visible or present in the mainstream circles of discourse but would nonetheless give pause to some of the popular prognosticators of our times if they were. That's because the social sciences, as they are normally practiced, are great at generating an abundance of models with complex explanatory power but of little to no practical, predictive value. By contrast, my heroes of the natural, hard sciences have kept busy forging practical, easy-to-understand economic models that offer clear, concise explanations with great predictive power. Many of the predictions these alternative theorists are arriving at are rather depressing, for sure, but they are at the same time fascinating and enlightening. And, as you will read, these women and men of alternative economic thinking have inspired me to take my own stab at explaining human economic outcomes via lessons from the animal kingdom and the natural order of things. Thus my obsession with ants, as indicated by this book's title.

So why ants, or an anthill? You thought we were discussing squirrels, right? Actually, the purpose of this text is to describe how lessons from nature, especially from the animal kingdom, particularly the insect world, hold powerful keys for far better understanding our own plight, including how we arrived at this particular point and where we may be headed next.

Do the entomologists also have their own radically different perspectives on human societal processes and future trends to share? Perhaps not, at least none that I have managed to locate as of this writing. But ants and termites are often used as apt, if imperfect, analogies for human civilization. If we are going to explore how nature forces humanity into certain patterns, then we may find value in exploring how this has already happened to other mass colony-type species on our planet. Ants provide as good an example as any.

Among the thousands of identified species of ants, we can find engineers, architects, farmers, and ranchers. They build cities, highways, bridges, landfills, and mass graves. They raid, conquer, stage coups, colonize, migrate, use tools, and enforce divisions of labor (sort of). They occasionally recycle. There's even a species of ant that practices a form of slavery. And they have absolutely no earthly idea why they are doing any of this and no capacity to willfully change any of these behaviors. There is also great redundancy within ant colonies, meaning few if any individual ants are ever missed should they disappear or get killed, rendering the vast majority of individual colony members useless and the ant cities amply capable of continuing to hum along nicely without them, even without scores of them. Did you also know that ant colonies can often suffer from massive levels of unemployment?

Much of the above probably sounds very familiar to you, I'd bet.

Ant colonies have evolved immense complexities without any conscious deliberation or cognitive calibration. Who's to say that we are any different? Why did we fan out across the world rather than just stay put in Africa, where we were born and where the weather is so much better suited to our hairless frames anyway? Why did we transition from hunter-gatherers to farmers, and then from farmers to urbanites? Why did we clump ourselves together and build huge ugly cities in the first place instead of dispersing ourselves out more evenly across the habitable landscapes? We idealize the quiet, rural lifestyle; yet we crowd ourselves into noisy, stressful cities anyway, just like the ants do. Is there any fundamental reason for this? No vote was ever taken, and it's not as if some committee somewhere made these decisions for us and then forced our compliance. It just sort of happened.

The naturalists, as it turns out, may help to explain why this all transpired the way it did, as you may have already guessed. They also possess powerful insights into our modern-day global economy and what most vexes it.

What's occupying these fringe scientists' minds of late? Well, to paraphrase the legendary rock band Aerosmith: there's something wrong with the world today, and we don't quite know what it is.

We do know that the world economy is slowing down—gradually, for sure, but a downward trend is unmistakable, as has been recognized by the World Bank, United Nations, International Monetary Fund, Organisation for Economic Co-operation and Development, and many, many other knowledgeable observers.

Technology continues to advance, for sure, but not at anything close to the pace at which it did from the end of World War II to about the 1970s or 1980s. In some ways technological innovation has even taken obvious and huge steps backward—for instance, there are no more moon landings, and the loss of the Concorde means slower air travel rather than the faster intercontinental connections that many predicted would have emerged by now. I'm still waiting for Ronald Reagan's space plane to take me from New York City to Tokyo in three hours—more than thirty-five years have passed since he first promised to deliver to us all a space plane, and the actual travel time is still closer to thirteen hours.

Fertility rates and average birthrates are in decline nearly everywhere, in both developing and developed nations. In fact, several advanced nations and China are poised to witness very large, significant population declines in coming years as their societies age and numbers of births fail to match numbers of deaths. Of course, we've already heard the usual explanations from the talking heads and activists espousing their own pet theories behind this one particular obsession of modern media: declining birthrates are the result of a lack of government-mandated paid parental family leave, we are told by some, or a consequence of rising income inequality, others argue. Some pin the blame on lapsing religiosity—get the people back to church, and they'll start making big families again, or so that argument goes. What if they're all wrong?

The talking heads also speak often of inequality, arguably the most popular topic among economists ever since a few recent national elections didn't exactly go as they would have preferred. Their initial diagnosis is correct: inequality is real, and it is getting worse, as is income stagnation, and yet the cost of living in many societies continues to climb higher. Thus a frustrated middle class is quite obviously losing ground economically, causing much of it to lash out politically as livelihoods continue to be squeezed. It is true that, worldwide, inequality is widening, not narrowing, as the world's financial experts promised would be the result of globalization, but again, why? Do the sociologist, business pundits, and economists we rely upon so much to inform us of why this is all occurring have the correct explanations, or are they missing the forest for the trees? Is it, as Thomas Piketty famously explained in his global best seller *Capital in the Twenty-First Century*, that inequality is a function of the rate of return on capital outpacing economic growth? Or is rising inequality an expression of something even more fundamental at work, something even molecular perhaps?

What do my heroes of the physical sciences have to say on these subjects? They say the answers lie in nature, clues hidden in the biological and behavioral comings and goings of the animal kingdom or sometimes in the motion of atoms and molecules in empty space. Declining net energy return on investment explains precipitously slowing economic growth and wage stagnation, they propose. Birthrates are collapsing because of density dependence. Global inequality is rising as a natural consequence of the laws of thermodynamics and something called "the rule of thirds." Gibberish for many of you now, perhaps, but by the end of this read, you'll have a much better appreciation for the foundational ideas underlying these alternative explanations and for why the natural sciences potentially hold within them far more powerful insights into humanity's present plight, and its inevitable future, than the social ones.

The predictions espoused by these alternative avenues of thinking will surprise you, and unfortunately in some cases they'll concern you as well. They may, in fact, even anger you. For instance, some of these radically different perspectives propose that a transition from fossil fuels to renewable-energy technologies will bring global economic growth to a standstill rather than providing that "shot in the arm" that many proponents of renewable-energy technologies insist they'll deliver. The rebel thinkers predict that further globalization and international market liberalization will inevitably deliver further economic inequality in tandem, perhaps not between nations but certainly within them, and that the world's political leaders will be incapable of preventing this outcome. Meanwhile, with a view through a different lens, some alternative laboratories of knowledge suggest that ongoing urbanization and net immigration will in fact exacerbate falling human fertility, rather than resolve it, thereby pulling population growth rates down ever lower, possibly even to negative rates, leading to eventual population contraction. And it gets worse: the laws of nature they point to as explanation for this supposed inevitability strongly suggest that not only is average human fertility set to continue its long, steady decline but average human life expectancy may begin falling as well in the coming years, perhaps soon enough for many of us to witness this change.

Perhaps most troubling, but equally fascinating: if these natural scientists are even half right and the fate of humanity is indeed governed by some fundamental laws of nature, then these rules could dictate that humanity is ultimately not in control of its own fate. In other words, if these alternative explanations from the hard, natural sciences turn out to be the correct ones, then we, collectively, may be no more capable of reversing the trends apparent in our economy/ecology than a herd of elk can control how they collectively forage or breed or evolve over the course of generations. If the ants and termites can't alter the future paths of their societies, can we?

Is humanity in control of its own fate? There are some reasons to doubt this, but these reasons, which I'm about to illuminate for you, are no cause for despair. Quite the opposite, for knowledge is power, and if we are to have any hope of reversing or at least adjusting to the most unpleasant consequences of our economy/ecology, then we will need to lean on a much broader swath of knowledge than is currently generated within the halls of central banks or university economics departments.

I'll repeat the question again: Can humanity determine its own fate? The natural scientists offer this answer: not necessarily, and not if some rules of nature hold as strongly for us as they do for animals, molecules, vegetables, and minerals. Or ants. We will explore their reasoning here.

Again, this all begs the question "What is the human economy/ecology?" Is it a carefully engineered thing bestowed upon us by a wise elite or by skilled technocrats? Or is it perhaps instead the inevitable consequence of long-established patterns in nature that equally determined the fate of all life on Earth, including squirrels and, yes, ants. If so, then this is useful, empowering information, and we shouldn't shy away from this possibility or refuse to stare it in the face.

An original working title that I had in mind for this book was *The Ant Farm*. The phrase conjures up an image of hundreds of tiny ants scurrying and tunneling and going about their daily chores while trapped in a glass box, unaware of their fate and lacking any escape route even if they were, as a dispassionate observer watches from above with mild interest, perhaps even with amusement. It's also an image and idea I borrowed from a scene in a movie I rather enjoyed.

The film is the 2005 American-German occult fiction piece titled *Constantine*, starring Keanu Reeves as the title character and Rachel Weisz as Angela Dodson. The scene in question has the two seated at a street diner about to enjoy a meal, while Constantine explains to Angela how he managed to find himself trapped between the cynical, uncaring forces of heaven and hell. She attempts to offer some helpful words of solace and encouragement—that it may be all part of God's plan, as she puts it. Constantine's reply is rather short and curt: "God's a kid with an ant farm, lady. He's not planning anything."

I

DENSITY DEPENDENCE

Where Have All the Children Gone?

It's in the news again as I write these words—a horrible, perplexing problem that has some governments alarmed and others up in arms, a major dilemma that's threatening to derail whole economies and set nations back decades, perhaps even centuries. If something isn't done immediately to reverse it, then there will be dire consequences for everyone, or so we are incessantly warned. Just what is this impending disaster? You've almost certainly already read about it on multiple occasions.

"Global fertility crisis has governments scrambling for an answer," declares an October 31, 2019, headline from Bloomberg. More recently, a July 15, 2020, health article by *BBC News* raises alarm over the "jaw-dropping global crash in children being born." "Korea's ultra-low birth rate needs state of emergency declaration," warns a lawmaker, as reported in the October 21, 2019, issue of the *Korea Biomedical Review.* The *Wall Street Journal* concludes that South Korea's famously low birthrates will eventually threaten the nation's security. "South Korea is having fewer babies; soon it will have fewer soldiers," the *Journal* warns darkly in a November 6, 2019, piece. Is Pyongyang destined to conquer the entire Korean peninsula, as this alarmist headline implies?

It's also deemed a crisis in the United States. "America's sex recession could lead to an economic depression," frets CNBC in an October 25, 2019, opinion piece. The author blames immature men and online pornography, along with artificial intelligence technology, a rather amusing assertion in my opinion.

Media panic over declining fertility and birth rates seems an almost seasonally recurring theme, but much of it is inspired by annual reporting by the Atlanta-based U.S. Centers for Disease Control and Prevention (CDC). Lately it seems that each new CDC update on the annual number of births recorded in the United States is ever more dire than the previous year's report, arguably justifying the intense media interest, such as this May 15, 2019, item from National Public Radio: "U.S. births fell to a 32-year low in 2018; CDC says birthrate is in record slump."

So where have all the children gone?

Despite their short attention spans, you tend to see members of the press return to this theme either following the release of fresh statistics by the CDC, or after publication of a report by some separate independent research group, or in the springtime. Perhaps that's because spring is the traditional season of fertility due to the obvious uptick in reproduction in nature (for instance, annual Easter holiday practices in the United States are clearly derived from ancient pagan spring rites featuring fertility symbols such as eggs, rabbits, baby chicks, etc.) or because that's generally when the CDC and a specialized agency of the United Nations tend to issue fresh population-trend reports. Wherever the inspiration comes from, musings about birthrates and trend lines in human fertility have never been more popular, or so it seems.

Some dark corners call it a "demographic winter" or the "baby bust" or what have you. However they phrase it, news reporters broadcasting about it and authors writing on the question generally agree that a seemingly intractable decline in the birthrates of a wide variety of nations threatens the very global economic fabric and general future prosperity. Some are celebrating this trend, for sure—frequently environmentalists with very legitimate concerns about the pressures that human population brings to bear on ecosystems. A declining human population would help to mitigate against global warming somewhat, for example. But those voices are far and away drowned out by the economists, experts, and talking heads thundering away with their warnings about the grave and inevitable consequences in store unless governments act to get a handle on this so-called problem.

A writer in the *New York Times* recently dubbed it "the end of babies."[1] A very cheery headline with no hyperbole whatsoever, for sure.

The following is a just a small sampling of some additional recent hyperventilating punditry on this subject, to further drive home my point.

"A birthrate crisis would require a whole new mindset on growth," pleads commentator Larry Elliot in a headline in the *Guardian* (March 31, 2019).[2] Here, he reviews a new book with the upbeat title *Empty Planet* and recommends that nations facing demographic headwinds both embrace immigration and learn to measure happiness by metrics other than gross domestic product (GDP) growth rates.

"China lawmakers urge freeing up family planning as birth rates plunge," writes a Reuters reporter (March 12, 2019).[3] He goes on to explain how China's despots are now reaping what they've sown by enforcing a draconian one-child policy on their population for decades. With some provinces now experiencing apparent population declines, a government in Beijing most disinclined toward freedom of expression and representative rule of law is now urging its citizens to have more babies, except the citizens aren't listening, at least not yet.

"Earth's population may start to fall from 2040. Does it matter?" asks the *Irish Times* (March 7, 2019).[4] The article reports on new findings by researchers disputing figures by the United Nations indicating that the number of people on the planet could exceed 11 billion by the end of this century (today the world's human population stands at around 7.7 billion). By factoring in plummeting birthrates, the article notes, dissenters to the UN's view argue that the worldwide human population will never reach 9 billion. Despite predictions of continual population increases out to 2100 by the forecasters at the UN Department of Economic and Social Affairs, many of us may actually live long enough to experience the world's population steadily and progressively decreasing, according to these contrarian views. I've come across a similar view in a Deutsche Bank report published a few years back; in it, an analyst predicts the global human population will stabilize at somewhere between 8 and 9 billion people.[5]

Those are among the more neutral, nuanced assessments found in the news. Other items that pop up in any casual online search for "birth rates" or "fertility rates" lean heavily in partisan directions. For example, liberal corners insist that birthrate declines are evidence that immigration levels and the welfare state must be greatly expanded, while conservative outlets forcefully argue that the trend is proof that abortion must be outlawed everywhere and that people should get themselves back to church as soon as possible. Whatever a particular story's supposed neutral stance or obvious political bent, the falling birthrates trend is mainly reported on in a negative light.

For instance, a "demographic shock" looms for South Korea's economy (*Korea Times*, March 28, 2019).[6] Meanwhile, in the United States, babies are being born at levels below what's necessary to maintain the population level, threatening the Medicare and Social Security systems (*CNN*, January 10, 2019).[7] And we're now informed that a perpetually low birthrate is providing political ammunition for right-leaning politicians in Italy heading into elections, sowing societal divisions (*New York Times*, March 27, 2019).[8] And the list goes on.

A casual perusal will turn up hundreds of reports approaching this very same topic in myriad different ways, all published fairly recently. There are certainly more to come. They exist for good reason. The central theme behind all of them is accurate and well documented.

But to return to my earlier question: Where have all the children gone? In fact, you needn't look far. Despite dour press reports such as those noted above, our world is still plenty blessed with children, even in places that are supposedly suffering from far too few of them. Take Japan, for instance, where I live.

Here I encounter children every day: walking to school, riding the bus, biking around the neighborhood, or walking their dogs. Former next-door neighbors of mine actually had five children of their own, something quite rare anywhere in the developed world these days, let alone in Japan. There was even a TV show that occasionally highlighted the busy lives of large Japanese families. One couple featured had produced eleven children.

So children haven't disappeared from Japan completely, of course, but there is no doubt that this nation's reputation for a dearth of childbirth is well deserved. Couples here simply don't have nearly as many children as they used to on average, and a rising number of individuals and couples go through their entire lives without having any children (including my current neighbors, an older couple living across the street). Schools in more rural districts are indeed closing for a lack of students.

The Japanese are by no means alone, though they are something of pioneers of this trend. South Korea, Germany, Italy, Singapore, France, Spain, China, and many, many more countries are now following Japan's lead. The United States is, as well, though the U.S. population is poised to continue expanding for some time. And the news media are fully aware of this rising trend, as shown in the headlines above.

Indeed, as you are reading this, it might be about that time of year for this peculiar seasonal obsession of the media to take hold and bubble up once again. Birthrate angst drives eyeballs and clicks, after all. Of course it would. In the United States, Europe, East Asia, Latin America, and, increasingly, portions of the Middle East, parts of Africa, and beyond, women and couples are indeed waiting longer to have children and are having fewer of them when they do. It's a fascinating phenomenon that should pique anyone's curiosity. A rising "childfree" movement even sees millions of healthy, happy, highly capable adults eschewing child rearing entirely, forming online communities to lend moral support to one another and to vent their frustration over the occasional snarky comments they hear from irascible family members none too pleased with their life choices.

For nations with "top-heavy" demographics, or a greater percentage of old than young people, this trend in lower numbers of births will eventually translate into either extremely slow population growth (United States, Australia, United Kingdom), population stagnation (France, Germany, possibly Canada), or inevitable population decline (Italy, Russia, Japan, Spain, South Korea, and others) because fewer women are entering their most fertile years compared to the numbers of women leaving them, as demographers explain. Though slow moving, these demographic shifts won't go unnoticed, especially by economists and economic statisticians. For instance, Japan, South Korea, Taiwan, Spain, Portugal, Greece, and Italy combined represent a massive wealthy economy of more than 320 million consumers, equivalent to the United States, where the population is set to fall steadily in the coming years. Some reports believe that China's population is already moving in reverse and that the Middle Kingdom could see a population drop measured in the hundreds of millions over the coming decades. That forecast might explain the panic in some corners of Beijing.

So why is all this occurring in every developed nation on earth and among a large swath of middle-income and developing nations as well? A variety of explanations are proffered, including several that contradict one another or just don't jibe with evidence and patterns detectable elsewhere in the world.

An October 19, 2018, article by the *Washington Post* charts the decline in U.S. birthrates and surveys the controversy surrounding this topic. Birthrates have dropped sharply for all races and ethnicities in the United States, in both rural and urban counties, says the report.[9] In the article, one economist argues it's because maternity- and paternity-leave benefits in the United States are too stingy. The report also mentions the blame assigned by some corners to abortion and an assertion by at least one famous media personality that immigration is the culprit, as it holds down wages and thus makes otherwise eligible bachelors less attractive to the opposite sex.

I can assure you that all these above explanations are dead wrong. And, for the time being, I don't need to cite any scientists or academic journals to explain how I know this.

It is a fair criticism that the United States lacks sufficient legal guarantees of paid family leave and that it suffers from a dearth of other family-friendly policies and practices that no doubt greatly complicates the lives of would-be parents. This is not a cause of the falling U.S. birthrate, however. If it were, then how is one to explain extremely low and dropping fertility and birth rates in nations with very generous family, paternal, and maternal work-leave laws?

The evidence against this argument is found in national fertility rates, which measure the number of children born to an individual woman on average over her lifetime (the birthrate, by contrast, typically counts the average number of births per one thousand population, but both methods basically calculate the same thing, despite some online commentators' protestations to the contrary[10]).

For instance, recent UN statistics place America's total fertility rate at 1.886 (estimated average covering the period 2015 to 2020). Norway is celebrated for its pro-women, pro-family legal protections, yet that Scandinavian nation's total fertility rate is estimated at 1.827 by the UN's statisticians, slightly lower than America's. Denmark hosts arguably one of the most generous and thorough welfare systems in the world, including an extensive cradle-to-grave safety net and ample public spending on child day care facilities. The UN's total fertility rate for Denmark, though, is estimated at just 1.762 for the same period, again lower than the U.S. fertility rate. With numbers like these, one might make the opposite argument, that generous family leave laws are detrimental to fertility, but that's not true either.

Abortion? Let's debunk this canard straight off. In the United States abortion rates are falling steeply and *in tandem* with birthrates. If these two factors were negatively correlated, one would expect to see the trend lines moving in opposite directions, not both angling downward together—meaning a decline in abortions should be correlated with more births, not fewer of them. Abortion rates are dropping because the number of pregnancies is declining.

I don't know enough to say conclusively one way or another whether mass levels of immigration depress wages to the point where child rearing becomes too expensive for many couples to contemplate or whether this effect is overblown. I have encountered smatterings of evidence uncovered by some research pointing to a drag on wage growth from high levels of low-skilled immigration, and considering how supply and demand work, it stands to reason that a nearly unlimited, constant input of labor will keep demand for that labor well fed and the cost of this labor depressed. But this impact is identified at the lower end of the wage scale. The paradox of sliding birthrates, in the United States and beyond, is that the trend appears particularly concentrated among women and men found higher up the income ladder and in professions not greatly impacted by immigration. Can a television talking head really pin blame on immigration for the decision by married lawyers, accountants, college professors, nurses, and even journalists and other TV talking heads also to delay childbirth, to have fewer children than their parents did, or to forgo parenthood entirely? I think not.

Some blame global warming. That's not the reason why couples are forgoing childbirth either, even if a few individuals and couples occasionally tell credulous reporters that it is. This excuse makes for interesting headlines, for sure, but the reality behind the birth dearth is far more nuanced, as I'm about to explain to you.

So what's behind the trend?

Forget all the economists, demographers, politicians, bloggers, and clergymen who've chimed in on this topic innumerable times and will continue to do so. We've heard their theories, and they've all fallen flat.

Do you know whom we really need to hear from? Wildlife biologists.

* * *

You may have already run across the story of the St. Matthew Island caribou herd. If not, allow me to retell it here, just briefly. It is a classic in wildlife observation history, a legendary tale often serving as a lesson in the dangers of population booms and busts.

Once upon a time, in the final year of World War II, the U.S. military got it in its head to build a small observation station manned by only about a dozen or so men on an exceptionally isolated outpost in the Bering Sea. St. Matthew Island, Alaska, is about as remote as it gets, but American forces had recaptured nearby Attu Island only a few months before all this, so the region was of some strategic value, at least for a while. However, the men had barely settled into their icy new home before the United States won the war and the government quickly abandoned its designs on St. Matthew Island.

Everyone was ordered off St. Matthew Island, but those men left something behind: about two dozen head of caribou, put there just in case station personnel found themselves in want of something to eat.

An article in the *Anchorage Daily News* reiterates the tale well (January 16, 2010). As the story goes, time passed, and scientists grew curious about what had happened to these two-dozen caribou abandoned on that island. Thirteen years later two biologists from the University of Alaska, Fairbanks, paid a visit to find out. The herd had expanded exponentially to more than thirteen hundred head, according to their count. Six years later the caribou population on the island was discovered to have more than quadrupled, to more than six thousand animals.

Then something really bad happened to the St. Matthew Island caribou. A survey conducted three years later barely counted forty individuals still alive on the island. Today there are none. The population completely collapsed. [11]

For years this tale was presented as a lesson in the pitfalls of invasive-species incursions, a stark example of what happens when a large mammal species lands in an ecosystem in which it encounters no natural predators but faces the threat of resource scarcity as its numbers increase year by year. The survey team that counted six thousand caribou wandering the island also noted the severe beating the island's already sparse vegetation was taking, as the news article recalls. Indeed, the island's caribou population was facing the grim prospect of mass starvation, and many of the animals likely succumbed in this way. More recent studies now argue that a particularly nasty winter storm might have killed off the vast majority of them, as the Alaska newspaper report notes. No one really knows for sure.

I'm more interested in the lesson that St. Matthew Island presents with regard to exponential population growth. In less than twenty years the island's caribou population had exploded from a few dozen individual members to more than six thousand, then crashed to nothing. How is this possible? And why isn't this sort of thing more common?

Scientists may look to the St. Matthew Island caribou story as emblematic of the ever-shifting balance between biotic potential versus the key environmental constraints enforced by both limiting factors and decimating factors and how the never-ending dance of these three forces of nature permit animal populations to either boom, bust, stabilize, or stagnate.

On earth, in the natural environment, wildlife scientists every day monitor and document the constant battle among these three forces. Most of the time no one dominates the other, and a gentle balance, or "détente," emerges after some generations pass. But in the end, limiting factors, or the fundamental constraints imposed by an environment, always win out, putting a ceiling on a species' population size. Occasionally a decimating factor can enter the scene like a bolt of lightning and do precisely what the name describes: decimate or depress population numbers in a big way, which is one theory for what happened to those poor caribou many years ago. In that instance, the theoretical decimating factor was a particularly harsh winter or a single really nasty winter storm. In natural circumstances (unlike what may have occurred on St. Matthew Island), decimating factors can actually sort of reinvigorate an animal population's biotic potential by placing the animals' numbers well below the limiting factor ceiling and allowing for a fresh spurt of population increase. But eventually the ceiling is reached again.

You are no doubt familiar with Thomas Malthus. He's the English philosopher who, toward the close of the 1700s, penned an essay predicting a grim future for humanity, replete with mass starvation and civil strife over food. As many of us were taught in high school, Malthus's theory pointed out that growth in agricultural food production on earth appeared to occur in a linear fashion, while the human population grew exponentially. Eventually human population growth would overtake humankind's capacity to expand the food supply, he predicted via relatively simple math. Thus, mass starvation and population collapse would be the inevitable outcome, Malthus hypothesized. His warning haunted the minds of scholars and policymakers for more than a hundred years until a humble agricultural scientist from the U.S. Midwest named Norman Bor-laug upended expectations by launching the Green Revolution.

Malthus's math on exponential population growth is solid and has been confirmed in nature over and over again through research, observa-tion, and experiment. Charts plotting the exponential growth of the hu-man population since the start of the Industrial Revolution serve as proof. Charles Darwin even noticed it as well, agreeing with Malthus that popu-lations do tend toward exponential expansion. But both of these great thinkers were apparently confounded by a reality that they and everyone else could see all around them every day: food resources were relatively limited, and population growth occurred exponentially, and yet animal populations in nature were not constantly prone to boom-and-bust cycles of rapid population increases followed by mass starvation and population collapse. Malthus likely glossed over or ultimately dismissed the implica-tions of this obvious contradiction, but it had to have weighed on his mind somewhat. If intelligent, capable humans are doomed to suffer such a dark fate, as he truly believed, then how do unthinking animal species figure out how to avoid this same horrible destiny? What loophole have they discovered?

To put it another way, why isn't the tale of the St. Matthew Island caribou far more common?

Exponential growth rates expose the great biotic potential of animal species, but this potential varies wildly among them. Some animals can lay thousands of eggs at once, theoretically arming those species with enormous biotic potential. The lionfish, for example, has become a prolif-ic invasive species in the Gulf of Mexico and Atlantic Ocean due in part to its high reproductive rate and lack of natural predators in these envi-ronments. Other animals reproduce very slowly, giving birth to only one offspring per cycle. The low reproductive rates of many whale species help explain why population recovery has been so slow ever since a global moratorium on commercial whaling was imposed in 1982. Species life expectancy matters as well; think of a cockroach versus a Galapagos tortoise. And there's a wide range between these extremes. But even animals with very low biologic reproductive capacity can experience ex-ponential rates of population expansion given enough time; an animal mother may only give birth to one child at a time, but she can still leave behind dozens of descendants so long as she can return to a fertile state and create more offspring over and over again over the course of her fertile life.

Darwin and Malthus had a fairly good understanding of this varied biotic potential in living organisms. However, they may have been less certain about the defining roles that limiting and decimating factors play (though Darwin clearly understood the role of natural limitations in driving evolution by natural selection). If Malthus knew what wildlife researchers know today, he might have felt compelled to either rewrite his famous essay or withdraw it from publication altogether.

Biotic potential and the key environmental constraints on it—limiting factors and decimating factors—are the yin and yang of the animal world. "Biotic potential describes a population's ability to grow over time through reproduction," as Clemson University scientist Greg Yarrow puts it. "A limiting factor is a basic requirement that is in short supply and that prevents or limits a particular wildlife population in an area from growing."[12] Biotic potential is fairly straightforward and is determined by an animal's biology. How many offspring can individual species members, usually female, engender at a time and in total over the course of their natural fertile lives? That number points to the species' biotic potential.

Limiting factors are often harder to pin down, even for trained scientists. Generally speaking, they involve things like the amount of food or fresh water available for animals to exploit in any given environment. But limiting factors can prove even subtler than this. For instance, are there enough tall trees with hollows for a species of owl to nest in and breed? Is a stream wide enough for a species of fish to navigate or find an adequate spot to lay its eggs? If the answer to both questions is no, then these negative attributes of the environments in which these animals find themselves very much limit their biotic potential as well, thus keeping their numbers low or preventing them from populating an area entirely. Like I said, subtle.

Then there are even more nuanced limiting factors, one of which I'll get to later.

Decimating factors are more obvious, usually pointing to things that contribute directly to animal mortality, including mass mortality. Predators keep herd numbers in check, for example. Disease sort of acts like a predator to keep a lid on things, as do disease-spreading parasites. Climate and weather can play a role, as the theory goes for the St. Matthew Island caribou. Natural disasters can also serve as decimating factors, from a hypothetical tsunami that could hit a species of lizard endemic to a single island to the meteor strike believed to have wiped out the dinosaurs more than sixty million years ago. In other words, these are much less subtle and involve the direct killing of animals. "Decimating factors can depress or reduce populations, but in most cases these factors do not control animal abundance," notes Yarrow,[13] although he's careful to add that, occasionally, this does happen. For instance, recently published research by a team at the Australian National University determined that a global amphibian pandemic caused by a fungal infection has been decimating frog species since at least the 1950s.[14]

So what determines an animal population's susceptibility to limiting factors (resource constraints, for instance) and decimating factors (like the frog-killing fungus)? In nature, ironically, one of a species' greatest strengths can also prove to be a significant weakness. Here, I'm thinking of the strength in numbers: population size and population density.

Consider various species of prairie chickens.

These are harmless, chicken-sized birds that nest on the ground and spend their days among the short grasses of the plains or coastal prairies of North America, hence the name (some prairie chicken species are critically endangered due to habitat losses). They are extremely difficult to spot, something that I can personally testify to. For starters, their plumage is the same color as the environment in which they live, providing them with excellent camouflage. But they also know how to hide and stay extremely still and quiet whenever they sense a potential threat.

Like farm chickens, prairie chickens can lay several eggs at one time and hatch multiple chicks. When their numbers are relatively low and their population is thinly spread out and evenly distributed, their biotic potential remains high—there is plenty of forage to be had since competition for food is relatively limited, diseases have fewer opportunities to spread through the population, and predators have a harder time spotting them the more thinly spread out they are. So their children can survive into adulthood, and their numbers can rise and rise as each generation produces more healthy offspring.

There are some advantages to a higher population figure, for sure, a big one being that it's easier for individual prairie chickens to find mates and spread their DNA. Closer concentrations of animals also better facilitate group cooperation, critical for survival in many species and a key driver of the "clumping" patterns wildlife researchers commonly find with population dispersal. But eventually a rising prairie chicken population becomes a liability, rendering its members more vulnerable to certain limiting and decimating factors with every annual increase. More birds on the ground mean that predators can spot or encounter them more easily, and diseases have an easier time spreading through a population. A higher number of prairie chickens also increases relative competition for food, nesting ground, mates, and so forth, the basic things that make prairie chicken life comfortable and at least somewhat predictable.

But here's the key: population size greatly matters in determining the degree of increased vulnerability to limiting and decimating factors, but not as much as population density does.

Say a population of some prairie chicken species consists of millions of birds (just hypothetically speaking). Despite a high population figure, if their average population density remains less than one bird per square kilometer, then they are still relatively safe from limiting and decimating factors. The population is still thinly enough spread out to be sufficiently capable of both avoiding predators and diseases and finding plenty of insects and seeds to satisfy the nutritional requirements of each individual prairie chicken, even if there are millions of them.

But if their population density rises to, say, five birds per square kilometer, or to ten, or thirty, or fifty, or one hundred, or even one thousand per square kilometer, then the situation starts to change rapidly. They become easier prey to spot. Communicable diseases have an easier time spreading. And competition increases for the limiting factors of available forage and shelter. For instance, it may become much harder for these prairie chickens to find suitable nesting grounds given the crowding and competition.

The above description is a simplification, but one designed to illustrate something. Indeed, population increases tend to be exponential: two become four, four become eight, eight become sixteen, sixteen become thirty-two, and on and on. Resource availability, on the other hand, is either fixed or expands in a nonexponential fashion. This is what alarmed Malthus, but he was only interested in gross population numbers. The dynamic changes completely at higher population densities—not just with higher numbers of individuals in a population but with the degree of this population's crowdedness. Predator success rises as prey becomes more densely populated. Diseases are more communicable and parasites more common. And the pressures that higher population density brings to bear on food and shelter acquisition, nesting, and the rearing of young can also have greater impact on animal behavior and biology.

This is what prevents a population, in normal circumstances, from growing exponentially out into infinity until all resources are exhausted and everyone starves to death. Growth is exponential at the outset, for sure, but as population density increases, population growth rates tend to slow precipitously until population numbers plateau.

Wildlife biologists have a useful term for this interesting phenomenon: density dependence. It's the loophole that nature discovered long before Malthus or Darwin ever had a chance to even ponder the possibility.

"A density-dependent factor acts in proportion to the number of animals per acre or square mile [or per square kilometer] within a population," as Colin Carpenter of the West Virginia Division of Natural Resources explains in a primer. "Natality and mortality often fluctuate with changes in density."[15]

Density dependence and how it works is still a topic of much debate in wildlife studies. There's disagreement over whether it dominates animal population patterns or is just one of several influencing factors; biologists also recognize the role that density-independent factors have to play as well. Nevertheless, the literature agrees that density dependence and density-dependent factors are very real and very important and are critical to understanding population dynamics in virtually all animal species.

What I've been describing above is how density dependence leads to fluctuations in mortality. Far more pertinent to our discussion is Carpenter's reference to fluctuations in animal natality—numbers of births relative to population size, otherwise known as the birthrate.

As mentioned above, greater population size and population density increase a species' exposure to decimating factors (such as predators and diseases) and limiting factors (like food or shelter availability). A sort of evolutionary arms race between predator and prey plays out over millions of years, but in the much shorter term the scales tend to tip to the predators' favor as prey population and population density rise—meaning that in ecosystems where predator-prey relationships exist, a rising prey population tends to encourage higher predator numbers, which then puts a check on the prey's numbers. That's just one example of how a fluctuation in mortality occurs via density dependence.

In terms of limiting factors, over time density dependence can lead an animal population to sort of prey upon itself, in a much less violent, more interesting way. If a population size rises and rises and relative resource availability more or less remains unchanged, then it's theoretically possible for food to run out for some individuals and for starvation to ensue. And, to be sure, this can and does occur. For instance, a study out of New Zealand determined that higher population density was responsible for rising juvenile deaths in the North Island saddleback, a species of bird. The researchers in that investigation hypothesize that the higher densities led to more juveniles ending up in subpar locations, where their parents were less capable of feeding and caring for them. [16]

But more commonly, long before that happens, a subtler, hidden limiting factor takes hold of both individual members and an entire animal population, and its effect is to push the birthrate in that particular species and population down to well below what it was during the period of exponential growth. The animal species doesn't literally prey on itself; rather, it responds to the pressures that greater population density brings to bear by preventing new generations from ever being born in the first place.

And thus density dependence solves Malthus's conundrum from both ends: greater population density encourages both greater mortality and lower natality, and the population levels off through rising death rates and falling fertility and birth rates, usually well below an environment's carrying capacity. "At high density, a trade-off between growth and reproduction delays the age of primiparity and often increases the cost of reproduction, decreasing both survival and future reproductive success of adult females," explains one team of researchers interested in how density dependence acts on larger herbivore species. They found that higher population densities lower both large herbivore body mass and large herbivore birthrates. [17]

A separate study has determined that higher population density in white-tailed deer leads to falling fertility among the deer and even to an increase in the average age at which females deliver their first offspring. "First-year females exhibited strong density dependence in fertility rate, with an almost complete cessation of reproduction at densities >30 deer/ square kilometer," said the authors. [18]

The power of density dependence has been seen to act on osprey populations as well. The osprey is a bird of prey, a raptor that loves to fish, common throughout the world. "A negative relationship between increasing density and breeding success has been found repeatedly in raptor species," the authors of that research paper inform their audience. [19]

These findings match results from similar studies that have uncovered, over and over again, a fairly strong negative correlation between population density and birthrates—meaning, in nature, increasing population densities appear to lead to decreasing fertility and fecundity and a lowering of a population's average birthrate. This in turn puts a check on population growth rates.

This effect has been witnessed and described time and time again for a wide variety of species. For instance, research has shown not only that higher population density will cause female rabbits to give birth to fewer litters but that the number of baby bunnies born per litter will also decline. Another common effect is to see the age of primiparity (birth of a first child) rise in species, limiting the potential number of babies that can be born to that female over her lifetime. And in some cases density dependence will see a growing number of individual members of a particular animal species failing to reproduce entirely. Yes, even in nature many animals will go childless throughout their entire lifespans.

Birds, reptiles, amphibians, land mammals large and small, you name it. Density dependence has even been demonstrated in laboratory settings to act on fruit flies.

That's right, fruit flies. Insects are bound by this law of nature as well, it appears.

Fruit fly fertility and fruit fly birthrates plummet as fruit fly colonies get really crowded, studies show. "Relative fecundities change with densities," concluded Stanford University researchers Andrew Clark and Marcus Feldman way back in an early-1980s research study, confirming other studies they referenced showing that "differences between isogenic lines in intrinsic growth rate decrease when populations are crowded." [20]

I bet you see where I'm going with this.

The natural power of density dependence takes hold as population density increases or as an animal's world gets more crowded. Its effect is to dramatically lower animal and even insect fertility and birth rates. This occurs with deer, rabbits, birds of prey, fruit flies, reptiles, and amphibians—and apparently with another species you and I are all too familiar with: Homo sapiens.

* * *

To be sure, our species has defied the odds and delayed the effects of density dependence for a very long time, and our 7.7 billion–strong population is proof positive of this. Our crowding has not made us more susceptible to predators, since we have no natural predators on this planet anymore and haven't for quite some time. And though millions of us have succumbed to communicable disease pandemics in the past, including diseases spread by parasites, more lately modern medicine has been winning out and preventing the huge loss of life witnessed during the black plague of the Middle Ages.[21] High rates of human infant mortality have been more or less defeated thanks to the same medical revolution, and the Green Revolution now yields to us far more food than Malthus or anyone from his days could have ever imagined—too much, in fact.

However, human fertility and natality have been, without question, falling, and falling in line with what's documented in other species. And there is powerful evidence that high rates of human population density are responsible for the trend, not immigration, abortion, inadequate maternity-leave laws, or any other single factor often pointed to. We can't blame global warming either.

The idea that density dependence might explain falling human fertility and birth rates first occurred to me during graduate university studies in ecosystem science, where I was initially introduced to the concept and its central importance in wildlife population dynamics. At the time I figured it was highly unlikely that I was the first to stumble upon this possibility. And I was correct—some other scholars far more skilled than I have explored this line of thinking as well, only there are a lot less of them than I would have otherwise imagined.

The earliest scientific reference I can find to the connection between density dependence in the animal kingdom and falling human birthrates is contained in a 2002 paper penned by Wolfgang Lutz and Ren Qiang for the Royal Society of London. In this paper, the authors note a lack of discussion on the role that density dependence may be playing in plummeting human fertility in the demographic literature. Studies that do explore the birthrate question, including some very recently published ones, remain biased toward the assumption that the falling birthrate phenomenon seen everywhere is being driven by socioeconomic factors and is thus an issue for sociologists, social scientists, economists, and politicians to consider and work out. Though both demographers themselves, Lutz and Qiang suggest that the knowledge revealed by wildlife experts should perhaps be used to take a stab at this mystery anyway, though they propose this in very diplomatic language. "Since human fertility, especially under the condition of conscious family planning during the later parts of demographic transition, is seen primarily as socially determined, ecological factors (such as population density), which are prominent in animal ecology, have played little or no role in demographic analysis," write Lutz and Qiang. "However, population density need not operate only through direct biological mechanisms; perceived population density may also be an important psychological determinant of fertility."[22]

My translation: since higher population density drives animal fertility lower, perhaps it does so for humans as well, but in an entirely different manner, of course, as we're certainly not animals, after all.

Lutz and Qiang imply in those lines that, yes, density dependence causes biological changes in animals that work to drive fertility lower, but since humans choose to lower their fertility, the effect must be psychological, not biological. It's an optimistic if biased assumption, and unfortunately it is wrong: the effect density dependence has on lowering animal birthrates works both biologically and behaviorally in those affected wild animal species as well, and behavioral change is just another way to term psychological change. Plus, an emerging body of evidence shows density-dependent factors are lowering human fertility in direct biological ways too. So our species may not be all that special after all.

Lutz gets better at tossing aside his bias as a human being in later research and draws a more objective assessment in a 2006 follow-up study, for which he has enlisted the aid of a veterinary scientist. This time his team gathered data on the birthrates of 145 countries and tested it against estimated population densities. Density ratios are generally calculated as the number of individuals per square kilometer, though this too has to be adjusted for large nations, like Canada, Russia, and Australia, that contain vast empty spaces but also see very low birthrates due, most likely, to their extreme climates and consequent extreme rates of population concentration into a select few cities.[23]

What does this 2006 follow-up study by Lutz and colleagues unveil? "We find a consistent and significant negative relationship between human fertility and population density,"[24] they report. The correlation they uncover is unusually strong—as human population density increases, human fertility and birth rates fall in tandem. Density is the factor that beats out almost all others, including relative national wealth or GDP. "Population density is now the most important factor explaining the fertility level, with only female literacy coming close in significance,"[25] they conclude.

You may now be wondering, How? What about higher population densities pushing animal and human fertility and birth rates lower? What mechanism is at work?

Earlier I mentioned that higher densities introduce greater exposure to both decimating and limiting factors. We humans have defeated the decimating factors and most of the limiting factors as well, but not all of them.

There's one powerful limiting factor left. It's that subtler, more nuanced limiting factor that I alluded to earlier. Despite our best science and engineering, it's still lingering—an enormously important element that we and the rest of the animal kingdom cannot escape no matter how clever we may become. It leads to very real behavioral and biological changes and is the most likely culprit behind the delayed childbirth in crowded deer populations, smaller baby bunny litters, lower fertility in large herbivores, and later and lower numbers of births within the human community as well.

It's a little thing called "stress."

Obviously not everyone in science and academia agrees with this view, despite Lutz et al.'s groundbreaking discoveries. After all, other demographers and social scientists have ignored their 2006 study pretty much completely, and economists have paid almost zero attention to it.

For instance, a November 2018 published study from the University of Pennsylvania begins with a note that more than 12 percent of women in the United States "have reduced fertility and/or fecundity."[26] That's an astonishingly high number. The study presupposes that either environmental factors or socioeconomic factors, or both, are to blame and sets out to find proof. The word "density" appears nowhere in this paper; nor does "stress," except as a verb. The study instead finds links to GDP per capita, air pollution, and body mass index or BMI (basically obesity).

Higher GDP per capita is a popular but puzzling correlating statistic often pointed to, one that experts often cite without ever explaining the cause-effect relationship. What is it about *more* wealth that makes people want to have *fewer* children? Shouldn't it work in reverse? Indeed, recent surveys reported on in the media suggest that Americans by and large want to have more children than they actually give birth to—couples with one child say they would be happier with two, and those with two children express a desire for three. Survey participants cite their own household economic worries as reasons for forgoing the extra offspring, so shouldn't these responses throw entirely into question the classic "greater wealth leads to falling birthrates" theory?

Not to mention the fact that we are now witnessing falling fertility and birth rates in nations still far less developed than the United States with much lower GDP-per-capita levels. Proponents of the "wealth is to blame" theory need to explain to me data showing, without a doubt, that birthrates are falling in poorer countries in Africa, the Middle East, Latin America, and Central and Southeast Asia. Some nations in these regions now have nearly reached the ultralow birthrates seen in Europe and the wealthy corners of East Asia, despite much smaller GDP and GDP-per-capita levels. Cuba is but one excellent example. Yet GDP and other socioeconomic indicators still hold sway in this discussion, though the evidence pointing to density dependence as the strongest explanation is overwhelming.

Not all professional economists are completely blind to the possibility that population density is the strongest factor at work here (though the vast majority of them remain so). For instance, David de la Croix and Paula Gobbi, economists with the Catholic University of Louvain in Belgium, grew curious enough to explore how population density correlated with birthrates in just forty-four developing countries. They found precisely what you find in the rich world and what the Lutz study discovered: high population densities correlated very strongly with declining fertility and fecundity. In fact, their study revealed that people are following precisely the same patterns as the white-tailed deer discussed above. "Duration analysis shows that age at marriage and age at first birth both increase with density," the Belgian economists write.[27]

Population density is the ultimate culprit, because, among other reasons, higher population densities increase animal populations' exposure and susceptibility to decimating factors and limiting factors. For our species and for other animals, in the absence of major decimating factors—no serious predator threats or vulnerability to diseases or parasites—and when other limiting factors like potential food shortages are not immediately an issue, then the principal limiting factor at work on a population is stress.

Crowding increases the effort necessary to find adequate shelter. For "nesting"-type species (like ours) this means more time spent finding adequate nesting or breeding grounds, meaning more time spent laying the foundations deemed essential for starting families, however defined. Thus, this additional time and exertion ends up delaying coupling and thus childbirth, and this serves to limit the number of offspring born to an individual female over her lifetime. You'll recall media reports fretting over the phenomenon of young adults in many countries living with their parents for longer and longer periods, sometimes even into their thirties.

Crowding may even delay the act of procreation itself—ironically so, considering that it should make it easier for individuals to find mates. That is, unless the very abundance of choices in mating partners makes would-be mates more finicky and discriminating in their decisions, thus causing them to take a longer time to "settle down" and get to the business of creating the next generation.

But the pressures of a crowded environment undoubtedly induce stress, and this added stress is apparently negatively impacting biological fertility as well. And we cannot pin all the blame on females, not by a long shot. For example, there is clinical evidence that elevated stress levels lower human male fertility.[28] Studies have determined that average sperm counts in men have been steadily declining over the past decades. Some have been pinning the blame on pollution, but now you know the real answer. In fact, if you read recent medical literature on this topic, it appears that the world of medicine is finally starting to take very seriously the role that stress is playing in rising rates of human infertility. It's about time, given that the medical research establishment has spent so many years assuming that the relationship only works in reverse—that infertility induces stress, rather than the other way around.

Crowding increases the effort necessary to find adequate food, water, and other resources as well. An animal that spends more time finding food, shelter, and resources has less time to devote to other activities, including coupling, mating, and child rearing. In some species, as an earlier-mentioned study indicates, greater competition for food and resources can actually lead to lower average animal body mass, which is also important for fertility and for delivering healthy, viable offspring (but that doesn't appear to be the case for our species).

For humans, food is in abundance but money is not, and we do not require only food to make us fit mates or high-potential family material. The increased competition for remunerative employment paired with a higher cost of living in our crowded urban environments seems to be working for us in much the same way as increased competition for food and shelter (or nesting ground) works in the animal kingdom. The rat race is rather stressful, after all, isn't it?

Consider, if you will, how the world's most crowded nations are also among the least fertile and record the lowest numbers of births per woman.

The fertility rate at which a population can sustain its current size is generally considered to be 2.1 live births per woman, as reports often indicate. This figure assumes that both parents need to replace themselves for the population to grow, while allowing for occasional instances of child mortality. But crowding compels average fertility rates far below this figure. South Korea is only roughly the size of Ohio; yet its population is greater than California's at about 51.5 million (current figures put California's population at just under 40 million, significantly fewer). South Korea's government is now struggling to turn around one of the lowest fertility rates in the world: 1.323 live births per woman according to UN data. Taiwan, a mountainous nation about the size of New Jersey, is believed to hold the current world record for the lowest national fertility rate: about 1.22. Taiwan's population is approximately 23.5 million (New Jersey, America's most crowded state, is home to about 8 million people).

The Netherlands is considered the most densely populated nation in the European Union. The Dutch have a particularly low birthrate as a consequence: 1.75.

Then there's Japan, where I live, famous home of the baby bust. This nation is about the size of California but home to more than triple the population at roughly 126 million. And unlike California, Japan's territory is approximately 80 percent mountains, leaving the remaining 20 percent somewhat suitable for habitation and agriculture. The United Nations figures Japan's average fertility rate stands at around 1.478 today (all these figures may have changed slightly by the time you read this). Much of the world's press spent quite a number of years looking down on Japan and its demographic (and consequent economic) woes. Now it seems evident that Japan was merely the Ghost of Christmas Future for a wide swath of the globe, whether or not these nations boast higher or lower levels of immigration. Even Canada, with robust net immigration and 22 percent of its population foreign-born, is now witnessing rapidly slowing population growth on the basis of falling fertility and stubbornly low birthrates, a trend driven by the insane crowding of Canada's citizenry.[29]

Higher population density begets more crowding, which begets more stress, which leads to fewer children. Why are societal fertility and birth rates plummeting nearly everywhere, in almost every country, rich and poor? The answer is clearly that density dependence is acting on higher and higher human population densities.

I don't know about you, but I don't think we'll be needing Scooby Doo and the gang to help us solve this mystery anymore.

* * *

This is undoubtedly a sensitive topic. The choice to have many children, just a few children, one child, or none at all is deeply personal and the business of no one but the woman or couple in question. Many individuals and couples will also find themselves trying desperately to have children at varying stages in their lives only to ultimately fail. Indeed, it's a very touchy subject that will undoubtedly hurt feelings. That is not my intention, of course. But we must view and analyze the subject objectively if we are to better understand this overwhelming tendency toward fewer births, a trend being driven first and foremost by population density and the stress it induces. So allow me to share my own personal experience, if I may, and demonstrate how an outside observer might analyze my situation from a more objective, scientific point of view.

My wife and I have no children. For years I dismissed the possibility outright, mainly due to the difficulty I encountered with securing gainful employment that earned a solid middle-class income. I once fanned out more than two hundred copies of my resume to receive just one invitation to a job interview in response, and that was after acquiring my first master's degree. A high cost of living and aggressive competition in the job market convinced me to forgo family life entirely, but eventually things turned around for me financially, and my wife and I had something of a change of heart as we grew older. We ultimately decided to try for at least one child. But by then it was probably too late—our efforts to bear a son or daughter ultimately proved fruitless, and after a while we both decided to just forget the whole thing and focus on building a happy life together. Scores of perfectly happy and content childless couples walk the earth, after all, including some of our very own good friends and family members. So we determined to join them.

Looking at it all objectively, one can see crowding and stress at work in every step of my life mentioned above and how this crowding-induced stress contributed to my eventually not having any offspring of my own. Initially I was stressed about income potential and future career prospects. Later, I was stressed about affordable, adequate housing and meeting rising living expenses, a stress that I only brought upon myself since I willfully chose to work and live in large, densely populated, metropolitan urban settings ever since finishing high school (including New York City, where I met my wife). Marriage didn't immediately make things better either; my wife and I simply shared this stress together, at least for a short while. And by the time we acquired the confidence to try for children of our own, the stress of age on our bodies likely made that entire process much more difficult.

We are hardly alone. Just within our own immediate circle of friends and family, my wife and I can count among us, as fellow childless or childfree individuals, my uncle, her aunt and uncle, my cousin and his wife, her cousin, longtime friends of my parents who are married, two of our engaged friends, a married couple we are friends with, four single female friends, at least two single male friends, and several previous neighbors back in Texas and current neighbors here in Japan. They're all in the same boat as us. Multiply these stories by hundreds of millions of individuals and couples finding themselves in similar or even identical circumstances, and it is easy to see what's become of the world's birthrates.

It's as if overpopulation is a problem that fixes itself.

Let's reiterate.

Humans are mammals. Mammals are animals. Animals live on Earth. And having arisen from the minerals, waters, and atmosphere of and within the bounds of the planet, animals living on Earth are constrained to certain rules of nature for which there are no appeals, pardons, or exceptions so long as an animal wishes to continue its survival, however temporary. They must all—especially the largest animals—eat and hydrate. Most must sleep, or at the very least rest at some point to conserve energy. They must expel waste. And they must inhale atmosphere to use energy from the element oxygen to metabolize water and food in accordance with molecular DNA instructions, all in order to rebuild and replace their bodies' constantly dying cells.

Animal populations, collectively, are also constrained by other certain laws of nature.

All things being equal, organisms on our planet tend to reproduce exponentially. Left to their own devices, species' populations will grow gradually until they reach an inflection point and then rapidly expand, adding an exponential number of new members with each new generation. But in nature this does not continue on indefinitely; if it did, the oceans and land would be filled to the brim, overflowing with critters. There are checks on population sizes, usually brought about by decimating factors (predators, diseases, parasites, the occasional asteroid or volcano) and limiting factors (finite resources, habitat, stress). Though an animal's population initially grows exponentially, eventually these checks on population growth increase as the population gets markedly crowded, until the death rate comes to exceed or match the birthrate and the population size reaches a more stable level or falls to whatever number that environment will sustain.

Wildlife ecologists call this phenomenon *density dependence*, a term that refers to the various ways in which the exponential growth patterns of a species' population comes to an abrupt halt as animal population density increases. Outside checks on population typically involve, as mentioned above, predation, disease, or parasitic illness, which reduce individual fitness and life expectancy. Absent these, an animal population achieves its equilibrium size through a different, most interesting means: rapidly diminished fertility. In other words, the birthrates plunge, sometimes dramatically.

This phenomenon has been witnessed time and time again, in a wide range of animals. As an animal population reaches the peak of its environment's carrying capacity, should no diseases or predators be available to keep the population size in check, that population responds by simply having far fewer children, putting itself in check that way. The stress induced by competitive pressures that come from living in an overcrowded environment is a key factor driving individual members to lower their reproduction levels, birthing fewer offspring via both physiological and behavioral (meaning, for humans, psychological) changes.

The human population has grown exponentially roughly from the time people started to extract and exploit fossil fuels in very large quantities. From an estimated size of approximately one billion individuals at the start of the nineteenth century, the world's population now exceeds 7.7

billion. Enormous advances in health care, vastly improved efficiency in the production and distribution of food and goods, and an extreme concentration of populations into cities and metropolitan areas (a phenomenon otherwise known as *urbanization*) have accompanied this rapid rise in population.

Now humanity's fertility and birth rates are plummeting to below replacement levels in many countries. And falling fertility is not a phenomenon confined to only rich nations. The defining features of the countries where this is happening, instead, seem to be overall national population density, especially urban population density, and rates of urbanization. This explains why plummeting birthrates are being witnessed not only in places like the United States, Japan, and Sweden but also in places such as Cuba, Iran, and Pakistan. And even vast nations with smaller populations relative to their land masses are not exempt from overcrowding and density-dependence-driven fertility and birth rate declines. Canada and Australia are both immense and much less populated than the United States, yet Australia and Canada are among the world's most crowded countries anyway, with the vast majority of their residents concentrated in just a handful of major metropolitan areas. Consequently Australia and Canada demonstrate fertility rates well below that magic figure of 2.1 live births per woman—even lower than America's, in fact, in the case of Canada.

Humans became dispersed from Africa and then clustered themselves into villages, then towns, then cities, and eventually megacities like Tokyo, London, Seoul, and New York. As a city's population density grows, competition for income-generating opportunities, shelter, and a host of other necessities of modern life intensifies as well, as is expressed by the rising cost of living in these urban environments. Additionally, competition from urban spaces may be what's pulling down fertility in more rural parts of the planet, especially in an increasingly interconnected global economy.

The premium for space rises. Competition for employment and even education grows fiercer. Life becomes more expensive. And newer generations being raised in these densely populated, hypercompetitive, expensive environments appear to be responding to the added stress and anxiety induced by life in these modern crowded realities by having far fewer children than past generations. There is strong evidence that the effect is biological, as well, and not just behavioral. Crowding-induced stress appears to be lowering both male and female biologic fertility, meaning our species' very biotic potential.

If this explanation holds true, then the cause is not higher educational attainment, lapsing religiosity, inadequate government support for families, or any other socioeconomic explanation commentators popularly point to as primarily to blame for falling birthrates globally; rather it is simple population density. And there's probably absolutely nothing that we can do about it either. Thus, most national efforts to encourage younger generations to have more children for the sake of economic growth and national stature have so far failed. Government initiatives and campaigns that may appear initially successful are also destined for ultimate failure over time. Density dependence has proven far stronger.

Prominent economists are increasingly expressing alarm at a possible forthcoming dearth of young workers and consumers in major economies worldwide as demographic forces shape the global economy. Yet, alarmed as they may be, there simply may be nothing that governments in Italy, Germany, Singapore, Japan, South Korea, Argentina, Iran, Cuba, Vietnam, China, Denmark, or even the United States can do to encourage higher birthrates. The natural law of density dependence may simply mean that achieving higher birthrates in highly urbanized societies with densely populated human settlements is an impossibility. Immigration may even exacerbate things if the net result is only elevated population densities, which push fertility and birth rates ever lower. Immigrants are smart, after all, and can be expected to flock to the cities for the same exact reasons the native population did so.[30]

Well, there is perhaps one way to resolve the matter. If population density is the ultimate answer behind plummeting birthrates (and it is), then the solution could be to just reverse it—to reverse urbanization and force a nation's population to spread itself more thinly and evenly throughout the countryside. Yup, simply reverse urbanization, and the problem could be resolved.

Good luck with that.

It's been tried before, and I don't mean Khmer Rouge style. One leg of Japan's efforts to boost fertility is an ad campaign encouraging younger Japanese to escape their cities and move to the countryside instead. The government is doing this only because it's noticed birthrates for rural Japanese are higher, but the initiative is failing anyway. Although Japan as a whole continues to lose population, the populations of Tokyo and some other metropolitan areas (Sapporo in particular) continue to rise.

Either way, I would advise governments not to attempt to reverse urbanization or to do anything at all about this so-called global fertility problem. Yes, you read that right. Though pundits and reporters everywhere are calling this a "crisis" and screaming for public policies aimed at reversing the great birth dearth, I'm bucking that trend here and will instead advise that politicians and bureaucrats across the planet do absolutely nothing at all to try to influence the numbers of births occurring among the world's population. Zero. Zip. Nada.

That's right. Frankly, governments should altogether stop trying to influence birthrates or the number of children couples choose to bear or not bear in their societies. For starters, it doesn't work. In addition, there already exists a long, sordid, and failed history of government meddling in maternity and family planning, and the activity always reflects the particular economic fetishes or concerns of the ruling apparatus at the time. Governments and their abetting economists attempting this everywhere have always had their own selfish interests in mind and never considered the interests or concerns of the population at large as important.

The most famous example of this is, of course, China's draconian one-child policy. Begun around 1980 because the authorities in Beijing feared China had too many people already and that its population was growing too rapidly, enforcement measures famously included a brutal campaign of forced sterilization and subtler penalties like denying government services to couples who had more than one child. Now Beijing frets about too few people and births—recent news says the birthrate in China just hit its lowest point since the Communists killed their way to power more than seventy years ago.

And it may surprise you to learn that the government in Tokyo once feared that Japan's population was far too high and rising too rapidly, in the 1920s and 1930s, when the nation's population was half what it is today. Back then the government responded by enticing thousands of Japanese to emigrate away from their homeland, particularly to Latin America. Thanks to Tokyo's lies, these migrants quickly met with extreme hardship and poverty before eventually finding their footing. This government emigration campaign has echoes today in the form of the vast Japanese diaspora found in places such as Brazil and Peru. Mass emigration from Europe in the late 1800s seemed to governments there to be a good thing at the time, but now authorities in Copenhagen are urging Danish citizens to "do it for Denmark!" in an embarrassing and ill-advised ad campaign featuring an attractive younger couple living out visual sexual innuendoes.

South Korea went through a population-control phase, too, when Seoul begged its citizens to stop having so many damn kids during the 1960s and 1970s. Now it's begging and pleading for them to do the exact opposite, but so far South Koreans are ignoring their government's panicked alarms and cries.

There is no worldwide "fertility crisis," as the pundits insist, including in some very recent and far-too-lengthy opinion pieces.[31] Populations rise, plateau, and fall, and it has always been and will always be this way. It's all perfectly natural, driven primarily by density dependence, and there is no escaping this reality, try as we might.

Perpetual growth and expansion in any system is an impossibility, especially in finite systems like the one we call Earth. It defies the laws of physics and nature and is thus an illusion. When economists insist that America's population must continue to rise and rise out to infinity, one must wonder at what U.S. population size their thirst for endless growth will finally be slaked: at four hundred million, five hundred million, two billion, perhaps five billion Americans? Luckily for us their dreams will never be realized—by 2100 the entire world's human population will probably be in the ballpark of eight to nine billion, and it will grow no further. I don't know what the peak U.S. population will be, and frankly I don't care. Know only this: if you do happen to have young children or newborns by the time you are reading these words, marvel at the fact that they may one day live in a future year in which the entire global human population has actually *fallen* by hundreds of millions or even a billion people, compared to only one or two decades prior to that time.

All species are bound by this unbreakable law of nature known as density dependence, no matter how clever they may be. Long before humans run out of food, fuel, or even minerals, we simply run out of space, a limitation that is self-imposed as our species' members continue to flock to the cities and crowd themselves voluntarily. So younger generations become stressed, causing fewer births to occur, putting total population size in check, in conformity with the laws of nature.

It's a powerful explanation for the fewer-babies phenomenon. It is the correct explanation. Yet there's a very good reason to root against density dependence as the force acting on the collapse in global human fertility and birth rates, because in nature density dependence works the overpopulation problem from both ends, as mentioned above.

The same pressures and stresses that induce young and mature animals to breed fewer offspring generally end up lowering the life expectancy of that population's older members as well. Death from old age is exceedingly rare in nature, after all, and if an environment of hypercompetitive resource acquisition doesn't bode well for fertility and natality, then it likely will lead to an earlier demise for aging animals that just can't keep up with their younger, fitter competitors.

This means that the density-dependence model strongly suggests that if average human birthrates are currently falling, then average human life expectancy is inevitably next. [32]

It's not difficult to imagine how this might transpire. Simple stress may be enough to wear down bodies in crowded environments, causing related diseases or health problems to become more common and thus giving rise to earlier deaths for adults. Crowding and the consequent decline in fertility may also spell doom for various nations' social safety nets, where they exist. Care for the old in our modern societies generally relies on taxing the labor of the young. If the population of the old rises while new youth are never born, then how do we ensure the elderly are cared for in the future? If there's no way to make that math work, then the ultimate short answer may be that society at large will simply fail to meet this challenge, resulting in higher numbers of elderly passing away earlier than might otherwise have been the case.

It's an exceedingly pessimistic forecast, one that I truly hope is proven wrong. We might see it first in Japan, South Korea, or Taiwan. Only time will tell.

2

EROI

Secular Stagnation in Nature

In 1966 U.S. economic growth, usually calculated as the annual percentage increase of the nation's gross domestic product, clocked in at 6.5 percent, according to data compiled by the World Bank. If that doesn't impress you, it should; authorities in China told the world that their country's economy expanded at about the same rate during 2018 (setting aside the question of whether or not we should believe them). China has boasted the world's fastest-growing major economy for some time now, though economic expansion there has now slowed to about 6 percent per year or less (the title of world's fastest-growing major economy often belongs to India).

China's story today was America's yesterday. Just as China's economy is now slowing down precipitously (possibly falling to the range of 4 percent growth per year by 2022, if not sooner), so did America's in an earlier time. Only the U.S. economy began its current deceleration in the later decades of the twentieth century. In 1976 U.S. GDP growth was estimated to be about 5.4 percent. In 1986, 3.5 percent. For 1996 the World Bank gives an economic growth rate for the United States of a bit under 3.8 percent, but by 2006 annual economic growth had fallen lower again, down to 2.86 percent or almost one full percentage point lower than the decade before—and this was during a raucous housing bubble that would famously implode only two years later.

U.S. economic growth is estimated to have been about 1.6 percent in 2016, per the World Bank's figures. Things are expected to get better, but not by much. As I write this, growth for 2019 is forecasted to come in at about 2.3 percent, according to the Federal Reserve Bank of Chicago. [1]

I mentioned this in the introduction to this book, but I'll say it here again: there's something wrong with the world today, and we don't quite know what it is.

Population growth is slowing as fertility and birth rates plunge nearly everywhere. We now have the explanation for why this is the case, the one you read about in Chapter 1: density dependence. Animal populations experience falling fertility at certain critical high-population densities, and we are part of the animal kingdom after all, whether we'd like to admit it or not.

But other mysteries of the modern human economy remain, in particular the mystery of our great global stagnation. What I'm referring to here is the measurable, verifiable, recorded and confirmed, and closely observed deceleration of the world's average annual economic expansion, with America's experience, as noted above, but one example of a wider trend.

The downward slide in annual U.S. GDP growth isn't linear—the World Bank's chart is a zigzag. The data shows a deep U.S. economic contraction in 1982, followed by a massive rebound, with GDP expanding by a whopping 7.2 percent in 1984. Yet, by 1985 U.S. economic growth returned to the 4 percent ballpark range common in the 1980s. And a longer downward slide is easily recognizable and hard to miss, even within the zigzag.

In the 1960s the U.S. economy expanded at annual rates commonly north of 6 percent. The 1970s were the 5 percent years. The 1980s and 1990s were 4 percent decades, more or less. The first decade of the 2000s will be remembered not as the wonder years but as the 3 percent years, despite the rise of the internet, Facebook, e-commerce, and all that noise.

Now we've just closed out the 2 percent annual growth decade. The 2020s will almost certainly be the decade of 1 percent annual U.S. economic growth figures, if the trend continues to hold. It won't be too much longer before the United States basically lands in the exact same boat as Germany and Japan, where economists find themselves giddy and in a fairly celebratory mood so long as GDP growth rates stay in positive territory, even if just a hair above 0.5 percent per year.

What explains this steady decline in annual U.S. GDP expansion? An even better question is this: Why is something similar happening throughout the world as a whole?

As the largest economy on the planet, the United States drives much of the global trend, but America's share of the global economy has been falling as Asian economies' shares rise. Yet a downward slide is also evident in the World Bank's historic national accounts data for annual worldwide GDP growth. It mimics the American narrative: in 1964 average annual global GDP expanded by almost 6.7 percent; in 1976, by 5.3 percent; in 1984, by 4.5 percent; and in 1997, by 3.7 percent; then it rebounded to about 4.4 percent growth by 2000, almost certainly thanks to China's rapid state-driven economic expansion, but continued its slide lower shortly after, down to less than 4.3 percent by 2010.

The International Monetary Fund (IMF) earlier estimated that the 2019 global economic expansion would clock in at a "still sluggish" 3.2 percent and hoped for a rebound to 3.5 percent during 2020.[2] The IMF was being far too optimistic, and it turns out that the economists there should have flipped those digits. More recent numbers paint a grimmer picture: the United Nations puts the world's GDP expansion during 2019 at just 2.3 percent—"the lowest rate since the global financial crisis of 2008–2009"[3] —and doesn't see things rebounding to anything close to the figure that the IMF is calling for. "Based on the assumption that potential setbacks will not materialize, a modest uptick in global growth to 2.5% is forecast for 2020," concludes the UN Department of Economic and Social Affairs.[4] From 6.7 percent in 1964 to 2.3 percent in 2019 represents a 65 percent deceleration of annual global economic expansion in five decades despite the huge world population growth that occurred during that same period.

Moody's Investors Service thinks the IMF's forecast for 2020 is far too optimistic as well. "Risks to credit conditions rise as the global slowdown takes hold," Moody's warned in its 2020 outlook for clients. Analysts at the ratings agency think the world's twenty largest economies will collectively expand by just 2.6 percent, if they're lucky. "Unfavorable demographic trends and low productivity growth in many countries increase the likelihood of a long-lasting period of weak growth," they wrote.[5]

Note Moody's reference to demographic trends, the subject of Chapter 1. Slowing population growth could be a factor for sure. Yet, at least in the United States, population growth has been fairly robust for most of those economically slowing decades. They include the baby boom years, after all. For the world as a whole, global population went from about 3.9 billion in 1970 to around 7.7 billion today, according to the United Nations Department of Economic and Social Affairs.

So, although the world's population has almost doubled since the 1970s, its economic growth has slowed by 50 percent over that same time period. And global economic growth continues to slow. Theories as to why this is occurring abound far and wide.

Quite a few observers pin the blame on demographics, as noted above. Some of them blame general productivity declines, suggesting that education and training in society are lacking. Others point a finger at social inequality made worse by a plague of rapacious capitalism.

There's also the "secular stagnation" hypothesis, "secular" here describing long-lasting conditions, not to be confused with secular in the sense of "irreligious." First posited during the Great Depression and revived in more recent economic discussions, the secular-stagnation idea holds that the world's economy is entering a kind of psychological funk in which demand is persistently weak despite low levels of unemployment and cheap borrowing costs. We'll discuss this concept in more detail later.

One interesting theory is a sort of take on the diminishing-returns hypothesis, positing that global technological innovations have been relatively lackluster as of late, dragging the world economy down as a consequence. "We have a steady stream of innovations but they're not as profound as ones in previous eras, and as a result we have a shortage of business investment because businesses don't find profitable, innovative activities to invest in." So explained Robert J. Gordon, a Northwestern University economist, during a conversation we had about his 2016 book *The Rise and Fall of American Growth: The U.S. Standard of Living since the Civil War.*

Gordon's theory behind the U.S. and global economic slowdown, as he discusses in his book, is that technological innovation is slowing, and thus GDP expansion is slowing with it. The premise of his argument is spot on: technological innovation and invention has slowed markedly, at least relative to prior years and eras. You may find this claim to be rather dubious given our current age of internet commerce, self-driving cars, whiz-bang "AI," and other marvels of modern gadgetry. But try to imagine things the following way.

Imagine a woman born in America in 1900. Her early childhood was largely a horse-and-buggy affair as far as transportation was concerned. Aviation meant a hot air balloon, and the fastest mode of travel available to her at the time was probably a steam-powered train.

By the 1910s she had seen her first automobile and perhaps ridden in one for the first time around then as well. The Wright brothers' innovations in heavier-than-air aviation, born in 1903, were by then rapidly spreading throughout the world economy, so quickly in fact that fixed-wing aerial combat played a major role in the world war that was raging in Europe shortly after our hypothetical female protagonist entered her teens.

By the time our fictional hero reached her twenties, automobiles were all but ubiquitous, especially on city streets. When she hit age twenty-seven, Charles Lindbergh made history by becoming the first person to pilot an aircraft across the Atlantic Ocean in his famous New York–to–Paris nonstop flight. Amelia Earhart accomplished the same feat in 1932, and she nearly succeeded in flying around the world but sadly disappeared in the central Pacific Ocean in 1937. That tragedy aside, our fictional protagonist spent her thirties in a fairly interesting time: aviation technology accelerated, and some additional remarkable innovations began to emerge as well, including radio detection and ranging (otherwise known as RADAR) and the first rudimentary computers.

By the time she turned forty years old, a second great global war had unfortunately broken out. Airplanes were then fitted to massive boats and even submarines. The first missiles were introduced to the battlefield. Yet the death and destruction didn't come to a complete close until the United States unleashed what is arguably humankind's most horrible invention: the nuclear bomb. Born at a time when experts were only beginning to conquer diseases and figure out heavier-than-air flight, our protagonist had now entered an era in which humans had split the atom, thereby achieving the power to level entire cities in a single blow. She now lived in far more interesting yet frightening times.

Two years following the first atomic bombing, Chuck Yeager broke the sound barrier, thus launching the jet aircraft age, which was off in earnest by the time our hero celebrated her fiftieth birthday. During her sixties, humans pierced the atmosphere altogether, orbited Earth, and

even walked on the moon. In her seventies the first probes landed successfully on Mars and Venus, and the bullet train was by then regularly
delivering millions of passengers to and from Tokyo and Osaka.

In her eighties the computing giants Apple and Microsoft emerged to
conquer the digital world. Cellular phone technology emerged around this
time as well, then the birth of the internet, just in time for our hero to
reach her nineties. High-speed wireless computing, GPS, and other mira-
cles of modern telecommunications were all gaining mainstream uptake
when our nameless fictional woman finally passed away at the ripe age of
one hundred.

On her deathbed this woman could reminisce about a remarkable lifetime of incredible first experiences. She went from horse-drawn transportation in her childhood to airplanes and automobiles by her teens and
twenties. Transatlantic travel, an amazing achievement in her thirties, was
now routine, as were transpacific passenger flights and faster-than-sound
travel. She witnessed the first flags planted at the South Pole and on
Mount Everest, but that was nothing compared to first flags placed by
human hands on the surface of the moon. Then in her old age she was
treated to the very first photographs taken from the surfaces of Venus and
Mars. The *Voyager* spacecraft graced her further with the first close-up
still pictures of the gas giants and outer planets, and she even witnessed
the initial construction of the International Space Station. Unfortunately
she wouldn't live long enough to see the first images of a black hole
existing at the center of a far-distant galaxy.

For me, the greatest technological changes that I can think of during
my lifetime, innovations that truly changed the way I live and breathe,
number just three: smart phones, laptop computers, and the internet. Cars
are pretty much the same today as they were when I was younger, at least
in form and function—they boast a few more electronic gadgets like
backup cameras, but for the most part changes in how they feel and
operate have been minor. Airplanes are almost exactly the same. Personally, I don't think that I or most of you reading these words right now will
ever chance to experience anything close to what this fictional woman in
my example did, and neither does Robert Gordon. And for him you don't
have to go all the way back to 1900 to realize this, as he explained to me.

"We had a complete transition in the way offices work from 1970
where we had a world of paper typewriters and file cabinets to the transition that first brought us memory typewriters. Then we got fax machines.

Then we got personal computers with spreadsheets, and we're processing software. And then finally in the 1990s came the internet with search engines and e-commerce. So we had a long transition from 1970 to about 2005 that brought us our current set of business practices that now are universal everywhere . . . and we're just not seeing further advances beyond that."

And try as I might, I failed to impress him with some of the more famous items on the modern-day techno-futurist's list of things we're all supposed to ooh and aah over and marvel at. "We've got lots of things going on: robots, artificial intelligence, 3D printing, and a gradual move toward self-driving vehicles," Gordon acknowledged. "But all of those things are happening very slowly. I go through my daily life watching the ways employees and businesses I visit operate, and I can't see a robot anywhere."

Thus his conclusion: economic growth in the past was driven by major leaps and bounds in technology, revolutions that changed the way we work, travel, communicate, and more. These revolutions are fewer and farther between these days and much less impressive or impactful than past leaps. Thus, the global economy is slowing.

It's a strong argument, one that could explain why growth has slowed so precipitously in some of the world's most advanced economies and manufacturing powerhouses. For instance, why has economic growth proven so anemic in Germany and Japan? Per Gordon's perspective, one could argue that the Germans and Japanese have already built their Towers of Babel, and they've simply run out of sky. It's very difficult to imagine how these two wealthy economies could become any more developed with existing technologies, and where advancements have been achieved, they've proven to be marginal improvements compared to past achievements.

Gordon's explanation for what lies behind the global growth funk isn't the only compelling one out there. It has a competitor popularly termed "secular stagnation," as mentioned earlier.

The London-based Centre for Economic Policy Research (CEPR) edited an entire book on the subject of secular stagnation featuring a variety of economic thinkers and contributors, including an essay by Gordon himself, in which he espouses his technology-is-to-blame thesis. He adds to this explanation various "structural headwinds" exacerbating the technological-innovation conundrum: demographics and stagnating population growth, a ceiling on educational attainment, rising income inequality, and massive public debts. All very compelling explanations in their own rights. [6]

But the book's opening chapter is reserved for the man who kicked-off the modern-day secular-stagnation debate in the first place: former U.S. Treasury secretary Lawrence Summers. In his contribution Summers notes the widening gap between potential forecasted GDP estimates and the actual figures recorded for the United States, Japan, and the European Union's "Eurozone," which are much lower since the 2008–2009 global financial crisis (for Japan the trend dates back to the famous bursting of Japan's speculation bubble in the mid-1990s). The weakness of the post-2009 recovery puzzles him and other economic observers. As witnessed in the early 1980s in the United States, sharp recessions are usually followed by robust rebounds, yet that hasn't been the case following the subprime mortgage bubble fiasco and the Wall Street–led global financial meltdown. The "recovery," as much as one could call it that, has been rather limp and anemic ever since.

Governments reacted to this crisis the same way Japan responded to its earlier implosion: dropping interest rates to record low levels with the intent of encouraging borrowing, discouraging savings, and thus pumping more cash into economies, ideally raising employment and growth in the process. Yet that result didn't materialize, as Summers pointed out when he first revived the idea of secular stagnation in a 2013 speech before the International Monetary Fund (the term was first coined by economist Alvin Hansen in the 1930s in his attempt to explain the stubborn Great Depression, which was only resolved by the outbreak of World War II). "Hence the possibility exists that no attainable interest rate will permit the balancing of saving and investment at full employment," Summers says in one essay. In plain English, this means that governments can drop interest rates as low as they like in the current environment, even making them zero or negative, and in this modern economic reality the growth they expect to occur will still stubbornly elude them. [7]

Summers sees many of the same structural issues at play as Gordon does. "Slower population and possibly technological growth means a reduction in the demand for new capital goods to equip new or more productive workers," he allows, agreeing that wealth concentrated into ever fewer hands poses an enormous growth challenge as well. "Rising inequality operates to raise the share of income going to those with a lower propensity to spend," he has pointed out.[8]

Summers also sees much psychology at work in secular stagnation. His thinking runs as follows: The 2008–2009 global economic crises left workers and companies skittish and less willing to invest or take risks while households are made to feel much more anxious and uncertain about their futures, a problem only compounded by the massive levels of indebtedness accumulated by governments in their failed decades-long pursuit of higher growth rates (along with household and private-sector debts, all at record highs). Lower interest rates will only encourage more debt, as would tax cuts, adding further to these anxieties and failing to deliver the growth fillip these measures are designed for. And the longer interest rates stay low, the more accustomed the population becomes to this low-rate atmosphere. Any attempts to raise rates again to rein in debt results in sharp pullbacks in investment, spending, and consequently economic growth.

In other words, Summers and other proponents of the modern secular-stagnation hypothesis see the world economy sinking ever more deeply into a sort of economic quicksand, whereby massively high debt burdens, ultralow interest rates, fewer technological breakthroughs, and weak labor force productivity and participation (and in some cases a shrinking labor pool), all coupled with rising income inequality and fear of the future, conspire to set the world's economy on a permanent stagnation momentum, and any attempts to break free from the trend may only exacerbate it instead. Companies hoard cash because of all that is listed above. Households are already stretched thin and don't spend proportionally more even when governments make it easier for them to borrow money. And where they have a vote, citizens often oppose further government stimulus via tax cuts or debt-fueled spending binges because they see their governments already drowning in red ink and worry about passing these burdens on to their children. Again, economists' use of the word "secular" doesn't mean nonreligious or nonspiritual stagnation; rather, it denotes the protracted, stubborn nature of the global growth slowdown "existing or continuing through ages or centuries," as a *Merriam-Webster Dictionary* definition helpfully explains.

Acolytes of the secular-stagnation hypothesis generally do a good job of backing up their theories, but they are fuzzier on proposed fixes. Summers ends his essay in the CEPR book by recommending even lower interest rates while urging governments everywhere to do something to encourage greater business confidence. He also suggests income redistribution, shifting more wealth to lower-income households, though he doesn't explain how this should be accomplished.

So behind the great global economic slowdown, we see plummeting fertility and birth rates, due, of course, to density dependence, the rule of nature whereby crowding induces stresses that compel sharply lower reproduction in animal populations. But we also see traditional economists leaning on more conventional explanations for our current global economic growth malaise. There are no more fantastic leaps in technological innovation, as one explanation holds. That a hangover effect from past recessions, especially the 2008–2009 crash, lingers is another assessment. Debt is too high, interest rates permanently low, and wealth concentrated in too few hands that don't spend or invest it enough—the secular-stagnation theory toolbox contains all these explanations.

Does there exist no other powerful theory behind this trend? Nature may offer one.

As it turns out, some of the alternative thinkers celebrated in these pages—including ecologists, biologists, and even a smattering of forestry experts and at least one mechanical engineer—have helpfully proffered another culprit for all of us to blame: energy. Or, to put it more precisely, another law of the natural world underpinning organism survival: energy return on investment (EROI).[9]

You will recall how earlier in this chapter I described the steady, unrelenting contraction in annual U.S. GDP growth since about the 1970s, from as high as 6.5 percent during the baby boom years to under 2 percent more recently. Interestingly enough, the history of U.S. domestic crude oil production shows a similar pattern; at least it did up until around 2009.

As reported by the U.S. Energy Information Administration (EIA), the statistical arm of the Department of Energy, in 1966 U.S. oil wells were producing on average about 8.3 million barrels of crude per day, and production continued rising to hit its first peak of 9.6 million barrels per day in 1970.

From there, oil output declines began taking hold. In 1976 total daily crude production was about 8.1 million barrels per day on average. A decade later U.S. domestic crude oil production had bounced up again but still landed well below the 1970s peak, with EIA putting 1986 output at almost 8.9 million barrels per day. Then, by 1996 daily crude production had slid much lower again, averaging less than 6.5 million barrels per day. Then it slid lower still. By 2006, just a few short years before the famous shale oil revolution took North Dakota and Texas by storm, daily U.S. domestic oil production was barely 5 million barrels per day, down from 9.6 million back in 1970.

Some astute observers have drawn a connection between the historic decline in U.S. economic expansion and a seemingly corresponding decline in U.S. domestic oil production. Perhaps the two are correlated, they've long argued. But if they are correlated, then what is to be made of a more recent phenomenon: as U.S. crude production has shot up in recent years thanks to the shale oil boom, flying well past that 1970 peak output level, America's economy has continued its long slowdown, despite this newfound crude oil bounty and domestic energy abundance. Why?

The data is there for all to see, published by EIA and the World Bank and easily cross-referenceable. In a single decade U.S. domestic crude oil production more than doubled from five million barrels per day in 2008 to nearly eleven million barrels per day in 2018. However, this boom apparently did U.S. GDP growth no good whatsoever. Even if one ignores the disaster of the 2008–2009 economic crash, you see no greater lift to overall economic growth rates from the modern U.S. shale oil boom whatsoever. While the economy grew by a respectable 3.8 percent in 2004, growth rates had slid down closer to 2.8 percent by 2018. The figure may clock in at just (or even below) 2 percent for 2019.

What's going on?

If sliding crude production and U.S. GDP expansion were positively correlated with one another in the earlier decades (just for argument's sake—I'm not saying the two figures definitely are correlated), then why didn't GDP expansion rebound with the roar of the shale oil years? Shouldn't we have witnessed at least an increase in annual economic growth rather than a continuation of the steady decline?

Practitioners of a field they've named "biophysical economics" might explain things this way: declining energy return on investment.

While U.S. oil production did indeed grow strongly from 2009 on, the effort it took the United States to get at this newfound energy abundance did not decrease in tandem. Rather, that effort probably increased instead—shale oil extraction requires much more drilling and wellhead activity per barrel of crude gained than was the case in the 1960s. Thus, a biophysical economist might argue that annual economic growth did not rebound whatsoever as the shale oil years roared but actually continued its long steady decline, because the United States has still been forced to exert much more energy to get at these new barrels of crude from 2009 until today compared to prior years of more robust GDP expansion. In other words, America's energy return on investment dropped lower for its crude industry compared to earlier decades, and thus its economy slowed with this decline in net EROI.

The same line of thinking could be employed to explain that earlier decline in GDP from the 1970s onward. As U.S. domestic crude production fell, U.S. oil demand continued to rise and rise. This gap was met with imports, first from mainly Mexico and Canada but eventually and increasingly from far-distant locales in the Middle East, especially Saudi Arabia. Thus, it can be argued that energy acquisition effort grew during these years as well alongside the U.S. oil import bill, and shipborne imports may account the most for this additional expended effort required to acquire these new barrels. Conventional domestic oil operations entail sending oil from a highly productive domestic well through pipelines in Texas to a refinery a few hundred miles away; importing crude from the Middle East, on the other hand, requires transportation overland, processing the crude at a port, loading it into a tanker, and then shipping it thousands of miles to another port, where it must be unloaded and transported farther to refineries in America. Whereas in the former case the oil travels dozens or hundreds of miles, in the latter example workers end up carrying it thousands of miles to satisfy America's thirst for energy.

To put it another way, importing crude from overseas involves many more steps, requiring more material and effort, regardless of the price per barrel of crude, as does extracting oil from tight rock and shale formations deep underground via continuous horizontal drilling and hydraulic fracturing.

In both cases alternative thinkers can see the forces of declining energy return on investment gradually weighing down on annual U.S. GDP growth.

In both cases—imports from Saudi Arabia and shale oil extraction through hydraulic fracturing—the United States has been forced to exert more and more energy to gain less and less of it in return. It can cost up to $3 more per barrel to import crude from Saudi Arabia versus Mexico or

$5 more per barrel than delivering crude from offshore Gulf of Mexico platforms, according to data compiled by the U.S. Energy Information Administration. So GDP expansion may consequently be suffering as a result.

First, imports from far-distant lands had to both offset domestic declines and fulfill the increase in U.S. crude demand. Then near-nonstop drilling and hydraulic fracturing activity took hold to get at those last locked barrels of crude still found in the continental United States, replacing a portion of imports in an environment of plateauing U.S. crude demand. The additional effort necessary to gain new energy perhaps results in a slowing down of the U.S. economy or at the very least does nothing to boost it. At least that's what one set of theorists believe. They see energy return on investment as paramount: as economies expend more and more energy to gain less and less of it in return, economic stagnation begins to ensue, say the biophysical economists.

* * *

As already mentioned in the introduction, ten years ago I got wind of a peculiar academic conference organized on the campus of the State University of New York College of Environmental Science and Forestry (SUNY-ESF) in Syracuse. Eventually I was invited to attend and cover it. It was hardly a major affair though; only the second gathering of its kind, it attracted maybe fifty participants at most, an eclectic mix of forestry scientists, evolutionary biologists, and a smattering of wildlife ecologists, experts in the hard environmental sciences for the most part.

Though of diverse academic backgrounds, these attendees of the second annual Biophysical Economics Conference were all top experts in their fields; bright, well-published tenured professors mixed in with accomplished postdocs and graduate students just launching their esteemed academic careers. But they all shared one thing in common: every woman and man there was absolutely convinced that the world economy is heading toward a precipice.

Their reasoning: declining EROI, or energy return on investment. It's a simple enough concept drawn from observations of nature.

Life, activity, momentum, what have you—all of it is only possible due to systems' or organisms' ability to obtain a surplus of energy over what is expended during various activities or operations. To put it even more simply, organism and system survival depends on acquiring a greater sum of energy than the energy expended to acquire that energy in the first place. It sounds a bit like circular reasoning, but it's not.

In northern Japan, bears gorge themselves on acorns and whatever other forage they can find in the autumn, then hibernate in a deep sleep throughout most of the winter months. As long as a bear can store up enough fat reserves in the fall for its metabolism to burn slowly and steadily during the hibernating months, with some left over to awaken, seek out, and obtain new food, then the bear will survive and thrive, ready to tackle the world anew in the spring. If forage during fall weeks proves insufficient, then the bear will starve to death. Oftentimes the result is somewhere in the middle—the bear survives winter but emerges from hibernation emaciated and in desperate need of new food.

The U.S. oil industry and U.S. economy work in much the same way, biophysical economists argue. To use an analogy, America is kind of like that squirrel in the introduction to this book. A slowing U.S. economy equates to this squirrel getting less and less energy intake for her trouble foraging for nuts each and every day. The squirrel exploits the easy-to-find nuts first, the morsels scattered just below and adjacent to her tree hollow, more than satisfying her needs for the summer and winter months with relatively little effort. Once these nuts are exhausted, however, the squirrel has to venture out beyond her tree, running farther out to neighboring trees and groves and to those stores of nuts, exerting more energy than was necessary in the past to carry this new bounty back home. Then even these nuts are exhausted. So the squirrel has to explore even farther beyond her normal range. Then farther still, and on and on, expending more and more energy every time to net less and less energy in the bargain.

The analogy is not a perfect one. Nuts and seeds are renewable resources—trees grow more of them each year so long as the weather is favorable. Oil is a nonrenewable resource. Our planet is actually generating new crude oil, but not at anything close to a rate that would be of interest to our species. Let's face it, none of us are going to wait around for millions of years for oil reserves to replenish themselves.

As with the squirrel and all things in nature, so goes American GDP, as the biophysical economics field argues. Despite the shale oil boom, America's EROI has been declining steadily since the 1960s, and thus the economy's real long-term potential growth rate has been declining with it. Or so the explanation goes.

The same pattern holds for colony species, including ants. In an environment of abundant and easily obtained resources, an ant colony can grow by leaps and bounds at the outset. But as the workers must venture farther out from the anthill for forage, the extra energy expended per individual lowers the net EROI of the colony, slowing or even stalling its growth. In some situations these ants may deem it necessary to relocate the colony entirely in pursuit of new resources.

For your average household, money can serve as a proxy for energy (some schools of thought actually link money and energy very closely). If a household spends less money each month than what it is capable of drawing in as income, then the family can keep a roof over its head, food on the table, and the heat and air-conditioning in working order, with perhaps a little left over for savings and entertainment. If funds fall short, then the household may respond by sharply reducing its expenditures or even selling some furniture, perhaps even the house itself in favor of a much smaller, more modest apartment—sort of "burning stored fat," if you will. If household income drops far below what's required even for a near-ascetic lifestyle, then that household may fall apart entirely.

In this simplified example, money merely represents the capacity of the household to acquire and expend new energy. For your standard, run-of-the-mill economist, household collapse occurs when funds run dry while expenses do not. A biophysical economist would see it as the household's expenditure of energy outpacing the homestead's ability to acquire new energy to expend, with the end result being bankruptcy.

Ten years ago participants at that SUNY-ESF biophysical economics conference saw world society and the world's economy as that household with oil as its money. Though globally oil production was increasing, oil demand was growing even faster, they feared. And they couldn't help but notice that crude output was falling in some jurisdictions, especially in the United States, where the modern-day oil industry was born. As went the United States, so would the world, they feared, with catastrophic consequences for the entire global economy.

Lead conference organizer Charles Hall, now retired, told me at the end of the conference proceedings that he and his colleagues could prove that oil resources were running out faster than new reserves could replace them. It's the same argument made famous by the "peak oil" theorists, only students of biophysical economics argue that oil will only be the first of many fossil energy stores and other resources to see catastrophic declines in EROI. Coal and natural gas would be next, Hall and others theorized, and no amount of technology could stop the fundamental law of EROI from taking hold of us all.

"Technology is in a race with depletion," Hall put it to me at the time. "And we can show empirically that depletion is winning."

Biophysical economists claim to have scientific evidence to back up this fundamental idea—that energy return on investment is what moves economies up or down—and many have published results of research that they say proves this phenomenon is at work.

A study by Jessica G. Lambert and colleauges, published in the journal *Energy Policy*, claims to have discovered a link between overall societal well-being and EROI, with the lead authors finding that EROI and energy use per capita "are as strong a statistical predictor as traditional economic indices" of wealth and overall economic well-being, household or otherwise.[10] In this study they share evidence that higher EROI leads to higher GDP, and not vice versa, arguing that a higher EROI depends on a society continually and reliably accessing abundant supplies of "high-quality energy." Through their assessment these scholars conclude that "having, or having access to, large quantities of high quality energy appears to contribute substantially to social well-being" because it leads to higher levels of GDP.[11]

A point of clarification may now be in order. By "high-quality energy" the authors are not favoring one energy source over another. The phrase doesn't reveal a biased preference by the researchers or a political opinion but rather stems from established quantitative and qualitative assessments that ended up labeling energy resources that are extremely energy dense, easy to store, and easy to transport as "high quality." Crude oil, coal, and natural gas fit this definition on all three counts. Wind and solar energy do not fit this definition, unfortunately, because you can't bottle up sunshine or a stiff breeze and carry it with you wherever you please. However, the electricity those technologies can produce is close to meeting the definition. Electricity is very energy dense after all (I invite you to stick your tongue on an electric fence if you doubt me) and fairly easy to transport, but it is generally difficult to store. To store oil, gas, and coal all one needs is a rudimentary container. Storing electricity requires complex chemistry and processed materials.

Hall saw the declining fortunes of the U.S. oil industry as the first sign of impending global doom. Of course, back then he and others in the biophysical economics community did not see the shale gas and shale oil revolutions barreling full speed ahead at us, with booming U.S. crude production just around the corner. Still, at the time their argument was extremely compelling and very hard to ignore, at least to me. Yet mainstream economists ignored it all the same.

For the keynote speakers, presenters, and active audience members at the SUNY-ESF biophysical economics conference, declining U.S. oil production prior to 2009 was of interest, but these scholars were far more interested in the falling EROI of the U.S. oil industry, or the fact that the oil industry found itself locked into a sort of treadmill to nowhere, stuck in a situation where it took more and more effort to gain new crude reserves at much less impressive volumes than a similar amount of effort would have netted these companies in the past.

First, easy-to-tap, relatively shallow wells in Pennsylvania were exploited and subsequently bled dry. Then, new land-based gushers in California and Texas were discovered and eventually fully exploited. This took the industry next to the Arctic in Alaska and then to the waters of the Gulf of Mexico. Today, if any new "gushers" of crude oil are found in the United States, they are likely to be discovered hundreds of miles off the coasts of Louisiana or Texas, trapped in fields deep underneath seabed, lying below seven thousand feet of ocean water, if not more. It's like that squirrel being forced to venture ever greater distances to find her needed winter forage.

Biophysical economists have parsed the pre-2009 U.S. oil industry's history into phases of declining U.S. oil-production EROI. In the 1930s they figure the industry's EROI sat at a ratio of about 100:1, meaning that drillers could gain a net one hundred barrels' worth of energy for every one barrel of oil burned to get at it. The industry's EROI ratio floated around that range for some time but began contracting sometime in the 1960s, right around the time when steady declines in U.S. GDP growth became evident, according to the theory. By the 1990s they figure the U.S. oil-production EROI had fallen steeply to about 36:1. In 2006 they put the oil industry's EROI at about 18:1, where it presumably stood at in 2008–2009 during the Great Recession. By 2013 researchers put the U.S. oil production EROI at about 10:1, just as the shale oil boom began to find its way to Texas by way of North Dakota. [12]

For new oil discovery, the picture they paint is much worse. Biophysical economists argue that U.S. oil explorers enjoyed an EROI on their exploration efforts of 1,000:1 way back in 1919—that they needed to only burn one barrel of oil's worth of energy to discover a thousand more recoverable barrels' worth to replace it. By the 2010s, with exploration dominated by either deepwater offshore drilling or onshore shale horizontal drilling and hydraulic fracturing, these researchers see new oil discovery EROI as having plummeted down to a ratio of just 5:1. [13]

In light of the shale oil boom reality, where crude production soared past the 1970 peak in 2018, where does America's oil EROI ratio stand today? I personally have no idea, though I doubt very much it has improved over the earlier 19:1 ratio estimated by biophysical economists, given what we know of the nature of shale crude extraction. Either way, U.S. GDP clearly didn't respond favorably to the shale oil boom. The data proves this.

That Lambert paper I mentioned above may point to the reason why: the researchers in that study find that economic well-being "appears to level off at [social] EROI values above 30:1," meaning "there is little or no additional improvement in societal well-being above these levels." [14] If true, this suggests that the great U.S. shale oil boom would have had to boost the nation's EROI by more than 50 percent (bringing EROI from 19:1 to at least above 30:1) in order to make its mark on the national GDP growth figures. Of course, no serious boost to GDP growth rates was ever registered, so in all likelihood the shale boom failed to lift America's EROI by that amount, if at all. And we should note that oil is just one source of total U.S. energy—oil, coal, natural gas, renewable energy, and so forth would all have to come together to greatly boost EROI in order to send U.S. economic growth on a tear again, according to the biophysical economists' theory. Obviously this didn't happen.

Back in the early days of the American oil industry, a company could drill a well at Spindletop near Beaumont, Texas, or farther west in the Midland-Odessa region, and it was like holding a license to print money—crude flowed at a large volume and steady rate for years. Eventually the company would notice a decline in output and could respond by either "stimulating" the well (adding pressure to encourage more crude to flow out of it or artificially pumping more out) or simply moving on to the next money-making hole in the ground.

In modern shale extraction the oil industry's drill bits are targeting not vast reserves of relatively easily accessible fluid crude oil found in vast fields of porous, spongy rock but rather molecules of oil so deeply intermingled within the rock that a lay observer would find it impossible to tell if there was any oil in it at all (believe me, I know, having held samples of this oil-bearing shale and tight rock in my hands).

Into this they drill, not only vertically, often as much as three miles down into the earth, but also horizontally, cutting a hole laterally through a formation to expose as much of the rock to the well bore as possible. Next, a tool is inserted to deliver a mix of water, chemicals, and sand (or an artificial "sand"), which is then blasted into the formation under extremely high pressure using sometimes dozens of diesel-powered pressure-pumping trucks working at the surface. The pressurized mixture forces hairline cracks to form in the rock, and the fluid mixture eventually leaves the sand in place to keep these cracks from closing in again under the extreme overlying pressures—thus the description "hydraulic fracturing." When all is said and done, molecules of oil flow under pressure into tiny spaces best measured in microns, gradually forming droplets as they coalesce into ever larger pore spaces, until eventually a steady flow of light, sweet fluid crude oil comes rushing out of the wellhead and heads onward to a refinery.

The flow doesn't last—within a year the volume of crude oil spewing from a horizontally drilled, hydraulically fractured well begins to decline precipitously, registering as an extremely sharp decline curve in the well owner's records. The company that owns the well then has two options: it can drill deeper into it, going farther laterally into the oil-bearing rock layer, and fracture it again for a second round of production (or third round, or fourth, or fifth); or it can move its drilling rig farther away to a more promising spot and start the process all over again. The result is a constant, never-ending process of drilling and fracturing to keep the pipelines filled and the refineries happy—and of course to keep the money flowing, salaries paid, and investors happy.

In a more recent conversation of ours, Professor Hall admitted that the U.S. shale oil boom threw him and his other colleagues in biophysical economics circles for a loop. They weren't expecting it (very few were, really), and he said it upset his and others' predictions of ever-sliding global EROI and eventual economic doom. Still, he regarded the shale oil story as a temporary bubble, a "Ponzi scheme," as he put it to me, that would eventually run its course, with the reality of EROI once again catching up to the industry and eventually the world.

But perhaps the rule of EROI never left, only paused or took on a different form.

The entire shale oil process, every step in it, is radically different from the way things used to work during the early days of the Texas oil rush. Considering this fact, it's easy to see how the shale boom might have failed to lift U.S. EROI. And even if EROI rose, it apparently did not improve to anything close to the critical 30:1 ratio posited in that earlier-mentioned study. The true EROI may have been a flatlining or even a modest decline since U.S. energy requirements have increased during the "shale gale."

However it turned out, the constant running of drilling rigs, pressure pumps, and trucks to and from well sites without question involves an expenditure of energy (diesel fuel and electricity) far above what used to be the case back when the first Spindletop gusher made Texas a global oil powerhouse. And, perhaps equally interestingly, through all this period of fracking and a doubling of U.S. domestic oil production, imports of foreign crude to the United States fell for sure, but not from where you might expect. Purchases from Venezuela and Africa were hit the most. Saudi Arabia and other Middle Eastern producers, meanwhile, kept shipping the same steady volumes of their crude to U.S. ports for most of the shale oil boom as they always had.

So overall energy-acquisition effort for America's energy needs probably didn't decline by all that much. If anything, this energy-acquisition effort actually increased, thus the ongoing steady fall of America's EROI quietly revealed in the anemic U.S. economic growth figures as America's economic slowdown continued apace.

And when America's shale oil boom eventually sputters out entirely (and it will), the U.S. economy will likely slow down even further as the country is forced to find other alternatives to the sort of easy-to-grab "high-quality" energy that fueled the twentieth-century economic boom that led America to become the global superpower that it is today. At least that's how proponents of the EROI theory see it.

Many of these experts think this shift will prove particularly painful. In other words "the decline in the provision of natural resources due to the future transition might be so large that it will significantly affect social welfare and economic growth," as Arizona State University scholar Adam Lampert put it in the journal *Nature Communications.*[15]

Though still ignored by mainstream economists, the importance of energy return on investment—the idea that EROI may be central and even possibly fundamental to economic growth and decline—has been slowly gaining greater interest in some corners. But Hall admits it's been an uphill battle for him and other EROI prophets. "No, we have not cracked the nut of conventional economics at all," he lamented, "although we also have many advocates from within and mostly without economics."

SUNY-ESF researchers Jessica Lambert, Charles Hall, Stephen Balogh, and other colleagues penned a lengthy assessment of EROI and its importance to economics for the United Kingdom Department for International Development. Published in October 2013, that report focused on the implications of dwindling EROI for much of the developing world that the UK delivers its foreign aid to. The authors argue that up to now the world has been oblivious to the natural law of EROI, free to ignore its implications thanks to fossil fuel abundance, but these days may fast be coming to an end. The global economic slowdown may be proof of this. They explain why they hold so strongly to this view. "At the societal level, declining EROI means that an increasing proportion of energy output must be diverted to attaining the energy needed to run an economy, leaving fewer discretionary funds available for 'non-essential' purchases which often drive growth and provide for social well-being," the paper explains at the outset. "Humans have tended to develop, and often deplete, the highest quality fuels first."

In other words, as a society is forced to expend more energy to simply gain and distribute new energy, proportionately less energy or effort is diverted to other economic activities, creating a drag on a society's overall economy.

Thus far we've been discussing oil-recovery efforts. What about the price of a barrel of oil, which ultimately determines the profitability of all this extraction activity?

Crude oil prices certainly matter. In the early days of "fracking," drillers hoped for a price of at least $75 to $80 per barrel to keep the party going. Back then crude traded in the $100-per-barrel range, so there was no problem until the 2014 oil price crash hit after the Organization of the Petroleum Exporting Countries declined to cut its own output to make space for American frackers in the market. Dozens of companies went bust, and thousands of oilfield workers lost their job as crude prices sank to the $30-per-barrel range. Then crude prices crept back up, mainly because traders noticed that crude demand was holding fairly steady, and the U.S. shale oil patch got its cost structures under control and kept the party largely continuing. Still, there is good evidence that companies relying mostly on shale oil fields to drive their businesses need a per-barrel price substantially better than $50 to keep it up for the long run. As I'm writing this, bankruptcies and layoffs in the shale oil patch are once again rising as better oil prices elude the industry and as these companies' investors lose patience.

But for EROI-focused thinkers the real central component isn't necessarily the price of energy or the cost of a barrel, or a kilowatt, or a joule. It is the total effort required to get at this energy in the first place.

The above authors argued to the United Kingdom's international development authorities that economic decline sets in as nations increasingly devote a larger proportion of overall economic activity to merely finding, producing, and distributing the primary energy necessary to run everything else in the economy. The more roughnecks, roustabouts, geologists, field managers, safety and environment managers, pipeline layers, truckers, lay workers, back-office staff, traders, bankers, lawyers, financers, butchers and bakers and candlestick makers needed to grab and move oil—and the more steel, rubber, wood, wool, cotton, concrete, copper, and other raw materials (and energy) needed for them to do their thing—the less there is left over to perform other kinds of work. And it's generally all the other stuff that drives most of the economic activity that registers in the GDP growth figures in the first place.

Actually, rising manpower requirements are likely not the culprit driving the oil industry's EROI lower and lower. I'm fairly certain that, at least for the United States, the industry's human resource needs have declined relatively over time, especially as a percentage of the greater economy's workforce. No, two other critical components necessary for oil drilling are acting to lower the oil patch's EROI: technology and capital. The industry invests in a lot more research and development (R& D) these days than in the past as it attempts to squeeze every last drop of crude out of the rock. And the oil and gas sector's capital requirements have shot through the roof compared to the industry's earlier era, even after accounting for inflation. A technically challenging offshore oil project can easily require a capital budget running into the tens of billions of dollars—no small change even for an international oil major. And shale horizontal drilling and hydraulic fracturing requires consistently open lines of credit to keep rigs running and pipelines filled. For hundreds of small to mid-sized fracking companies in the United States, the banks really do hold the cards.

So the frackers may have netted the United States much more oil, but if they accomplished this with way much more effort (money, technology, new R&D, etc.) above and beyond what was required in 1919 or even in the 1950s, then at best it's a wash in GDP terms, and at worst the industry's EROI is continuing to fall.

But the beginning of the broader trend is timed to the late 1960s and early 1970s, back when domestic oil production last peaked. This, say biophysical economists, is when the nation's EROI began its current long, steady fall downward. GDP expansion rates fell down with it. And something else has been in steady decline since then as well—namely, the share of wages paid to labor in the size of the overall American economy, or the wage share of GDP. Meanwhile the amount of energy consumed per capita in the United States has pretty much flatlined since then, neither rising nor falling by all that much year to year. This may be hard to believe, but the U.S. government's own statistics bear this out: a person living in America today consumes about as much energy per year as a person living in 1975 did.

The only thing that has enjoyed robust growth throughout this entire period is debt, both public and private. Debt burdens have exploded almost everywhere, in real and nominal terms and as a percentage of GDP. Household balance sheets are swimming in red ink, as everyone knows only too well. So are governments'. The U.S. federal government's debt had risen to an astonishing $22 trillion by early 2019, a new record high according to National Public Radio.[16] The World Bank recently warned that the debt burdens of developing countries have skyrocketed as well, climbing to an all-time high of $55 trillion by the end of 2018. The Bank put emerging economies' debt-to-GDP ratio at 168 percent, an increase of 54 percent over a period of just eight years. "The size, speed, and breadth of the latest debt wave should concern us all," warned World Bank president David Malpass upon the release of the new data.[17]

Carey King, a mechanical engineer by training and professor at the Energy Institute of the University of Texas, Austin, suspects rising debt burdens aided the continual (though slowing) expansion of the U.S. economy right about the time when the nation began overexploiting its easy-to-get, high-quality energy and found itself forced to rely more and more on the harder-to-get varieties, whether from imports, fracking, offshore Gulf of Mexico deepwater drilling, or what have you.

"My hypothesis is as you suggest," King confirmed when I posed the idea to him. "Resource constraints of the 1970s . . . forced us to change the rules of the economy, such as getting off the gold standard, repealing the Glass-Steagall Act. That led to and/or enabled rapid debt increases." It was either that or see very little to no GDP growth at all, Professor King argued. "The trade-off to have a growing GDP seemed to be increased debt and stagnant wages, or lower wage share."

Most subscribers to the biophysical economics narrative don't expect any salvation from a renewable-energy revolution anytime soon either, though many of them enthusiastically endorse a broad switch to alternative energy sources anyway. They are scientists, after all, and know fully well the physical processes underpinning global warming and the risks inherent in allowing greenhouse gases to continue accumulating in the atmosphere at such high rates, as is the case now.

On the off chance that some of you reading this now count yourselves among the climate science "skeptics" crowd and hold the opinion that industrial society can expel unlimited volumes of exhaust from tailpipes and smokestacks into the atmosphere to infinity with no consequences whatsoever for global environs, allow me one brief, feeble attempt to illuminate for you the reality.

Here goes: Earth's atmosphere is insulation, correct? Yes, it is. And what happens when you make an insulator thicker?

Yes, global warming is a major challenge, and wind power and solar photovoltaic (PV) electricity plants and installations are powerful tools for mitigating the damage of the fossil fuel age. But these energy sources are probably not the economic panacea that many of the most ardent proponents of renewable energy claim them to be, according to biophysical economists and the EROI model.

I'm referring here to those who hold that renewable electricity is the answer not only to climate change but to all the things that ail modern society and are subjects of this book: global economic stagnation, wage stagnation, and rising income inequality, among other socioeconomic conundrums. It's been argued frequently, in the United States and beyond, that abandoning fossil fuels in favor of an emissions-free wind- and solar-powered world will not only be good for the environment but provide the ultimate shot in the arm that our global economy needs right now. Research into energy sources and EROI suggests otherwise.

The reason is that the biophysical economics field does not qualify these renewable-energy technologies as "high-quality" energy sources. Again, this isn't because they have biased or uninformed views of what wind and solar can provide but rather because they've quantitatively compared wind and solar with all other popular forms of electricity generation—natural gas, coal, sometimes oil, nuclear, hydropower, geothermal, you name it—and have found again and again that wind and solar power demonstrate some of the lowest potential EROI of all generating sources, at least given existing technologies.

A lack of portability and difficulty with storage is one issue. Another part of the reason is the intermittency of wind and solar energy resources, for sure.

Solar power works best, of course, with clear skies at high noon in daylight and doesn't work at all during the evening hours. In most jurisdictions, peak demand on a power grid occurs sometime in late afternoon, beginning around 4 or 5 p.m., about when kids are coming home from school and when commuters are returning home from work but while the power is still on at the offices they've just left. This is right about the time when solar energy systems begin delivering less energy to a grid, not more.

Wind energy presents a more mixed picture, but it's well known that, at least in Texas, the wind-energy capital of the United States, wind-power systems are running at their most efficient in the middle of the night when the vast majority of the population is asleep and in need of very little of it. Wind energy overgeneration at night in Texas became such a problem for power companies that rate payers in the Dallas–Fort Worth metropolitan area were offered evening wind-generated electricity for free, just to give it all someplace to go.

Modern electricity grids depend heavily on power plants that can deliver a defined volume of electricity, kilowatts and megawatts that can be scheduled ahead of time to meet projected demand spikes. Solar and wind can't offer this flexibility, not yet anyway. That wind and solar may now be cheaper on a cost-per-kilowatt basis than fossil fuel generation in the United States and in many other jurisdictions is irrelevant. I could offer to sell you the cheapest laptop computer you'll ever find, but if I warn you that you can only use it in the daytime and only under a blue sky, would you still be interested? This is the very reason why developing Southeast Asian economies are assuming a paramount role for fossil fuels in their grids moving forward: they're trying to entice factory investment, and manufacturers demand consistent, reliable access to electricity twenty-four hours a day. Grid operators want power plants that can deliver predictable power and schedule their loads in advance ahead of heat waves, holidays, or maintenance work at other power plants. Wind farms and solar power arrays can't offer this flexibility. But intermittency and an inability to schedule beforehand when the wind blows and the sun shines isn't the only reason for the low EROI rating researchers have applied to these renewable-energy resources.

A thorough study into this question published in 2013 provides a case in point. In research led by the Institute for Solid-State Nuclear Physics in Berlin, scientists in Germany, Poland, and Canada set out to quantify and then compare and contrast the EROI values for wind, solar photovoltaic, solar thermal, coal, natural gas, hydropower, nuclear, and "biogas" (or biomass) energy generation. They even offered wind and solar sources a leg up in their assessments: their models assumed that wind energy and solar PV were "buffeted" with backup energy-storage systems, including grid-connected battery storage and even pumped hydro storage systems.[18] Yet even with this attempt to put all these energy sources on a more or less even keel, the study concludes, "nuclear, hydro, coal, and natural gas power systems (in this order) are one order of magnitude more effective than photovoltaics and wind power."[19]

This German-led study is actually rather critical of other biophysical economics papers that came before it—the authors take a hard view of some earlier work published by former SUNY-ESF professor Charles Hall, for example. Yet the team for this research arrived at largely the same conclusions as the broader biophysical economics field in general does. As a result of their analysis, they assigned wind and solar PV the lowest EROI values of the power-generation sources studied (biomass landed at the bottom with them), with or without storage backup systems. They reached these conclusions after doing their best to factor in the entire value chain of these energy technologies, from raw materials acquisition, to materials processing, to equipment and unit manufacturing, plant construction, operational performance (load per year), operational lifetime, maintenance, and decommissioning.

"Solar PV in Germany, even with the more effective roof installation and even when not taking the needed buffering (storage and over-capacities) into account, has an EROI far below the economic limit," the researchers concluded. "Wind energy seems to be above the economic limit but falls below when combined even with the most effective pump storage and even when installed at the German coast."[20]

A separate Finnish paper written and designed explicitly to advocate for solar and wind over carbon-heavy electricity generation was drafted with the authors apparently already fully understanding that these renewable-energy technologies demonstrate much lower EROIs than their fossil-fueled competitors. But the University of Helsinki research team argues that the world needs to go all-in on these technologies anyway, given the threat global warming poses. "Societies need to abandon fossil fuels because of their impact on the climate," these authors demand. "Because renewables have a lower EROI and different technical requirements, such as the need to build energy storage facilities, meeting current or growing levels of energy need in the next few decades with low-carbon solutions will be extremely difficult, if not impossible."[21] Hardly a ringing endorsement for renewables, and yet these authors' intention is to strongly advocate for a renewable-energy transition.

Impossible though it may be, this particular paper concludes that a rapid and thorough transition from fossil fuels to mainly wind and solar or other low-carbon energy sources is critically necessary, even if it entails considerable economic pain and sacrifice ahead given these energy sources' much lower EROI ratings. Thus, this transition must be forced into reality by governments as markets will not adequately create the incentives necessary for the general public to follow, the authors conclude. They believe the coming public resistance to this painful change will be fierce and that it should be countered—fought back, as it were—with government promises of greater environmental benefits and a good life, albeit a poorer one for many. "In rich countries, citizens would have less purchasing power than now, but it would be distributed more equally," the authors propose.[22]

Again, hardly ringing endorsements for a renewable-energy revolution, it may seem, but only from the point of view of EROI and what drives growth and rising standards of living in national economies. Such a shift, were it to occur, would undoubtedly be good from a climate standpoint, however. These experts are only saying that if you accept the central EROI thesis underpinning biophysical economics theories, you should not expect that switching out our gasoline-powered cars for wind-powered electric vehicles will bring about a new economic renaissance. Rather, biophysical economics thinkers are convinced that a shift toward renewables will still see the world economy continuing along that EROI downslide, running on a treadmill to nowhere as countries expend more and more energy to net less and less of it in return and thereby downshifting broader economic activity and sending GDP growth stalling. And they insist that they can prove it.

It all sounds rather ominous, but the biophysical economists don't know everything. They didn't see the shale oil boom coming, for instance.

Furthermore they are likely wrong, I believe, about looming declines in EROI for coal and natural gas, at least in the near term. Reserves of both remain abundant, almost ridiculously so, and these reserves can still be relatively easily exploited, especially if one looks outside the United States. But crude oil runs almost everything—you need crude oil to get at the coal and natural gas in the first place and to manufacture and deploy renewable-energy systems[23] —and new major oil discoveries are definitely getting harder and harder to make compared to the past, as any oil executive would confess.

And though they may still exist in abundance, these fossil energy resources are not infinitely abundant. Natural resources have been exhausted in the past by many societies, after all, followed by regional economic collapses. It can certainly happen again, especially with a resource like crude oil, which takes millions of years to form, unlike fish stocks or rainforests.

Still, it would be beautifully poetic if wind and solar were the answer to our great global economic funk. All energy is derived from the sun, after all, and wind and solar exploit this fact more directly than fossil fuels. But there are still numerous steps involved before humans can convert the sunshine and the breeze into something that you can use to power your houselights and television sets or to charge your smart phone. And apparently the overall effort involved in achieving this modern electrical feat is much, much higher than the effort it took to drill a hole into the ground in western Pennsylvania in the late 1800s and burn what came out of it. Thus, the switch from oil and coal to electric cars and renewable electricity will likely see the global economic slowdown continue, per biophysical economists and their EROI thesis. It may even exacerbate it, they say.

Either way the biophysical economists see the story ending the same: total economic collapse once EROI falls so low that far too much of the world's economic activity is devoted to simply acquiring and moving more energy, leaving little left to do much of anything else.

Our ultimate escape from this grim fate may lie in tapping an even more fundamental source of energy, purer and rawer than either the sun or the wind. What if we could skip the sunlight entirely and go straight to what generates those photons in the first place?

The world's largest experimental hydrogen fusion reactor is under construction in southern France. It is already on track to becoming the most complicated machine ever built by human hands, and probably the most expensive. Its builders are aiming to achieve the force that drives the entire solar system here on Earth: large-scale, controlled nuclear fusion reactions that net far more power than they take to generate. It may become the greatest energy source ever exploited by humankind, if they succeed. It's something physicists and governments have dreamed of for decades.

The International Thermonuclear Experimental Reactor (ITER) going up now near the small French town of Saint-Paul-lès-Durance is currently more than 65 percent complete and on track to achieving "first plasma" sometime by the end of 2025.[24] ITER representative Sabina Griffith told me in July 2019 that the world's first commercial-scale fusion power reactor will soon receive its core unit, the tokamak. She also warned that additional patience would be required of fusion power's enthusiasts, even after that first plasma milestone is cleared. "It will take another ten years until we reach full deuterium-tritium operations," Griffith explained.

Success is far from guaranteed.

Although physicists have managed to create fusion reactions under ultrahot conditions in controlled laboratory environments on numerous occasions, these have all been small-scale experiments. The ITER project is operating at a much more massive scale. The machine is to be fitted with the world's largest and most powerful electromagnets, powered by superconductors, all of which is required to generate a powerful magnetic field. This ultrastrong magnetic field will be necessary to contain a plasma core that will become so hot that no substance would be capable of touching it without vaporizing instantaneously (oven mitts would do you no good whatsoever). The fusion reactions are supposed to occur within this plasma core, hopefully generating far more net power than what was necessary to generate these fusion reactions in the first place.

It is fusion that powers the center of the sun, but we humans can't replicate the conditions at the sun's core because we can't reproduce the massive gravitational pressure that helps generate the fusion going on there in the first place. So the ITER core will have to operate at temperatures much hotter than the sun's core, about ten times hotter, in fact, or at around 150 million degrees Celsius, in order to compensate.

We should all be rooting for ITER's success. A nuclear fusion power plant would be the stuff energy return on investment dreams are made of—it could theoretically have an EROI that's completely off the charts, far beyond anything humans have exploited before, without producing any greenhouse gas emissions or dangerous levels of radiation.

Will the nuclear fusion visionaries behind ITER put an end to the doomsday prophecies of the biophysical economists? Again, only time will tell.

3

ENTROPY AND INEQUALITY

The Natural Disarray of Things

You have probably heard of Thomas Piketty, a famous French economist and professor at the Paris School of Economics. Piketty was made world famous by his best-selling book *Capital in the Twenty-First Century*, first published in French in 2013 with the English translation issued in 2014. The book's central topic is economic inequality, a much-discussed problem lately as economists, politicians, and the general public haven't failed to notice that the poor and middle classes in much of the developed (and some parts of the developing) world are finding themselves struggling like never before, while the rich see their wealth, prestige, and political power increase nicely year after year. Perceptions of inequality are seen to be driving recent political trends, including a rise of populism discerned by many a talking head on television news programs. Lately pundits like to pin the blame on inequality for election outcomes in the United States, the United Kingdom, and Latin America that haven't exactly gone as more left-leaning thinkers would have preferred.[1] I see some of these recent election results as merely a sign of the electorate venting its frustrations, opportunities for the working classes to deliver a giant middle finger to all these pundits, experts, and other successful urbanites who in earlier years seized the reins of power, insisted that they knew what was best for everyone else, and then forged ahead as they saw fit, leaping into globalization ass first while refusing to listen to any and all dissenting voices. But here I digress.

For a time quite a large number of these figures saw technology as at the heart of rising inequality in the United States. This was a fad for several months—reporters invented a fictitious "robot revolution" to explain the decline in manufacturing employment in America. The robots destroyed all the high-paying working-class jobs, we were told, not the decades of explicit U.S. government policy encouraging manufacturers to move abroad, where factory owners could take advantage of much cheaper labor and far weaker environmental laws and standards. Except if this were true, then all those shuttered, rusting old and abandoned factories in the Northeast and Midwest wouldn't be abandoned at all; they'd be filled with robots instead of humans. The U.S. media never bothered to explain this contradiction, unfortunately.

I can tell you from my own reporting on this topic that the United States is among the least robotized advanced economies in the world, not the most. Countries that do boast higher degrees of automation and utilization of industrial robotics generally see higher levels of employment in manufacturing and higher manufacturing wages too (Japan and Germany are but two excellent examples). This is because technology leads to productivity increases that allow industries higher profits and opportunities to expand, leading to rising employment. In the United States, 90 percent of industrial robots are used for automobile manufacturing alone, and automotive industry employment has consequently risen during this trend rather than declined.[2] Where automation has hit employment in the United States, it has mainly affected the white-collar workforce, not the blue-collar one, owing largely to the rise of the internet (which has hit journalism particularly hard) and things like tax-filing software and automated legal discovery applications used by law firms, reducing the need for accountants, lawyers, and paralegal services. But accountants and paralegals, among others in the white-collar workforce, tend to be fairly highly educated and have, for the most part (though not always), managed to find alternative employment at comparable pay levels.

So if we don't have the robots to blame for rising socioeconomic inequality, then what?

Piketty's central argument in his now famous book is that over time the world is seeing capital's rate of return rising faster than the overall rate of economic growth—meaning, people who control great wealth are seeing faster returns on their investments at a rate outpacing expansion of gross domestic product. For workers, wage increases tend to hinge on

GDP growth rates. Since GDP is growing more slowly (perhaps because of declining energy return on investment), while capital investments are enjoying relatively higher returns year after year, the share of global wealth going to the wealthy is consequently outpacing that which goes to the workers of the world, leading to higher income and wealth inequality, or so his explanation goes.

The book presents this thesis in basic mathematical terms as well. Let r represent the rate of return on capital investments (corporate profits due to shareholders, stock dividends, earned interest, the rents owners can charge, etc.), while g represents basic economic growth. Piketty's book says that rising wealth inequality is explained by a simple formula: $r > g$.[3]

I've read *Capital in the Twenty-First Century* cover to cover. It isn't exactly riveting, edge-of-your-seat reading, but it delivers its central point quite nicely. Piketty believes that $r > g$ is a fundamental feature of capitalism, not a mistake or some bug that can be worked out with a few tweaks here and there. Labor productivity and a resulting increase of labor income will not help either, he argues; nor can the problem be fixed by sudden, unexpected leaps in technological advancements. Even accounting for these, capital's gains will still far outpace labor's gains, causing a widening income and wealth inequality gap over time, according to his research and modeling.

He's careful to note that the concept as he's laid it out explains inequality gaps not among workers themselves—say, between computer scientists and fast-food restaurant employees—but rather between workers and managers/owners. But as the gap between the rich and the rest widens over time, the world can expect to see greater social discord and political instability and rising incidents of labor conflicts, including strikes, riots even, and other occasional bouts of violence, Piketty fears. He argues that the solution is higher taxation of wealth and government-driven redistribution, in whatever form that may take. "A market economy based on private property, if left to itself, contains powerful forces of convergence, associated in particular with the diffusion of knowledge and skills; but it also contains powerful forces of divergence, which are potentially threatening to democratic societies and to the values of social justice on which they are based," he warns.[4]

Is Thomas Piketty right?

It's just one idea, one explanation among many for the inequality dilemma, but it has had a big impact on the debate. Piketty's book has both fans and foes, many of whom have penned detailed arguments both for and against his ideas. Nevertheless *Capital in the Twenty-First Century* and the $r > g$ equation have made Thomas Piketty virtually a household name for a broad swath of the economics profession, academia, popular media, and members of the economics-curious general public. For sure, you've most likely already heard of Thomas Piketty.

I doubt very much the same can be said of Victor Yakovenko, a professor of physics at the University of Maryland, College Park. No, the vast majority of you reading this have never heard of him, but Yakovenko's theory behind the conundrum of rising global economic inequality is far more interesting and compelling than Piketty's.

Yakovenko was born in Donetsk, Ukraine, back when it was still part of the Soviet Union. He earned both an MS and a PhD in physics at schools in Moscow and fled the Eastern Bloc for good shortly after the Iron Curtain fell and the Soviet Union collapsed. His CV has him employed at Rutgers University by 1991 and then eventually at the University of Maryland starting around 1993. There he has remained, earning tenure by 2000 and then, in 2004, the title of full professor. He's also researched and taught abroad, holding visiting positions in France, Italy, and the United Kingdom. Yakovenko has even spent a few months at the Los Alamos National Laboratory in New Mexico.

Though he is an accomplished physicist by training and profession, lately Yakovenko's interests have drifted into economics. He's careful to note that he waited until after he secured academic tenure to pursue these additional hobbies—better safe than sorry, after all. But via this new path, Yakovenko has declared himself an acolyte of a new interdisciplinary field of study and training that calls itself "econophysics." Where the field of biophysical economics has a central, foundational concept in the form of EROI, the field of econophysics has one of its own as well. But before I tell you what that is, let's explore the problem of inequality and its potential consequences in a little bit more detail.

In early April 2019, the Organisation for Economic Co-operation and Development (OECD) declared wealth inequality and the "struggling middle class" to be an acute challenge that demanded immediate response by all OECD member governments. "The middle class has shrunk in most OECD countries as it has become more difficult for younger generations to make it to the middle class," the organization warned. "Across the OECD area, except for a few countries, middle incomes are barely higher today than they were ten years ago, increasing by just 0.3% per year, a third less than the average income of the richest 10%." Though middle incomes rose only barely to not at all, the cost of living has risen substantially. For instance, analysts there estimate that OECD member countries' housing prices have more than doubled since 1995, while average incomes increased by barely 25 percent over the same period. This and other factors are causing more and more middle-class earners to fall into poverty, the OECD secretariat warned in its report.[5]

Piketty and other renowned economists delved deeply into the topic of inequality again with their "2018 World Inequality Report." They find that the gap between the rich and everyone else began soaring around 1980 then leveled off somewhat with the 2008 global financial crisis. This report concludes, as the OECD did, that the middle class suffered the worst hit during this trend, experiencing stagnant or falling pay to meet rising costs of living.[6]

A 2016 study by the International Monetary Fund (IMF) focused on the phenomenon as it is occurring just in the United States. Here the IMF concludes that U.S. low- and middle-income households began experiencing income stagnation in the 1970s, while "real incomes of households in the highest brackets rose sharply during 1970–2000, though have not changed considerably since 2000." Explanations for the trend run the gamut, the IMF finds, with its economists and the academic literature they lean on pointing to "technological progress"—somewhat the opposite of Robert Gordon's theories (you'll recall Gordon's thesis that tech advancements are decelerating, not accelerating, as the IMF argues)—along with "declining unionization, taxation, international trade, education, immigration, household structure, and demographics."[7]

The IMF's findings that the trend in widening inequality in the United States (and perhaps globally) may have tapered off or ended entirely in 2000 partly inspired a recent contrarian take on the topic by *The Economist* magazine. Many argued that the IMF's findings are illusory as they don't take into account the much higher living costs and educational expenses today, compared to 2000, or the fact that workers have been forced to pay ever higher percentages of their incomes for housing while the wealthy have largely avoided this fate. Ignoring these points, in a pieced titled "Inequality Illusions," *The Economist* doubles down on the Fund's claim, finding little evidence that the top 1 percent of U.S. income earners pulled even further ahead of their compatriots since the turn of the new millennium. "In America, official data suggest that the same measure rose until 2000 and since then there has been volatility around a flat trend," the magazine says, essentially echoing the IMF. And even this existing inequality has been alleviated by expanded aid for the poor and fancy smart phones, among other innovations and gadgets, according to the author(s).[8] The entire piece is designed as a defense of capitalism and a call for minor tweaks around the edges, rather than a major overhaul, but I don't see these arguments gaining much currency with the modern electorate in democratic societies these days, where populations know well and good that lawmakers by a wide margin promote the interests of the rich first and those of the rest whenever they can get around to it (a political power slant that's been proven again and again in statistical social science research).

The experience of the broader general public in the United States mostly matches an explanation for the rise in inequality proposed by David Autor, David Dorn, and Gordon Hanson, a team of economists who turned their microscopes on the 2000 decision by the administration of President Bill Clinton to encourage China's entry into the World Trade Organization (WTO).

An op-ed penned at the time by the Economic Policy Institute (EPI), a think tank, predicted with stunning accuracy what would ultimately come of China's WTO accession, a Clinton pet project that Autor and team strongly believe paved the way for a certain populist, iconoclast billionaire's later election to the White House in 2016.[9] The WTO entry would lock in U.S. trade policies favorable to China (in one direction), EPI warned, thereby "setting the stage for rapidly rising trade deficits in the future that would severely depress employment in manufacturing, the sector most directly affected by trade."[10]

And there's more. "China's accession to the WTO would *also increase income inequality* in the U.S.," EPI predicted (the emphasis is mine).[11] EPI's 2000 warning proved so prescient that it makes me wonder if the institute happened to employ psychics at the time or perhaps controlled a time machine. Despite Clinton's bizarre insistence back then that the WTO-China deal would shrink the U.S. trade deficit with China, the trade deficit exploded instead, and U.S. manufacturing sectors shed millions of jobs with astonishing speed, almost certainly exacerbating income inequality.

But don't take my word for it; this is precisely what happened, according to research findings by economists Autor, Dorn, and Hanson. "Adjustment in local labor markets is remarkably slow, with wages and labor-force participation rates remaining depressed and unemployment rates remaining elevated for at least a full decade after the China trade shock commences," they wrote. "At the national level, employment has fallen in U.S. industries more exposed to import competition, as expected, but offsetting employment gains in other industries have yet to materialize."[12]

In other words, trade normalization with China and its WTO entry crushed U.S. manufacturing, as EPI warned it would. And despite the mainstream economists' insistence at the time, offsetting employment did not widely materialize as they promised, and where it eventually did, these changes meant formerly well-off workers were forced to take much lower-paying jobs in order to survive. Meanwhile, the cream of profits from exploiting cheaper Chinese labor and environmental standards flowed to factory owners and wealthy shareholders, thereby greatly widening the income gap in the United States. This is essentially the Autor et al. paper's take on the rising inequality phenomenon. It's just one explanation for the explosion in income and wealth inequality, among many others.

This "trade is to blame" proposition is the closest to physicist Victor Yakovenko's theory. His argument is rather different and unique, however.

For starters, Yakovenko sees the rise in inequality as a consequence of a greater openness within and between all countries, not just China, and impacting all nations on Earth, China included, because, as keen observers will tell you, the era of globalization has seen a dramatic decline in inequality between nations (particularly developed versus developing) but rising and exacerbating inequality within these same nations. There's a pretty good indication that the gap between the wealthiest and the poorest is as high as it's ever been in almost all developing countries, including formerly developing states like China, as this gap has also widened in the richer nations like the United States.

Is policy to blame? Were those anti-WTO protestors in Seattle in 1999 correct all along? Yakovenko, the contrarian self-described econophysicist, doesn't necessarily see things this way, which is what makes his alternative explanation so damn interesting.

Yakovenko thinks this rise in global inequality is perfectly natural and in many ways inevitable. In laying out his argument for what lies behind the inequality trend, he relies firmly on the rules of the natural world. In that vein his ideas mesh quite nicely with the earlier concepts introduced to you in this book: density dependence (the explanation for falling birthrates) and falling energy return on investment (one possible explanation for global economic stagnation), theories that place natural forces at the heart of the evolution of the human economy.

So why is economic inequality on the rise in much of the world? Because the odds favor this outcome, as Yakovenko would more or less explain it. "The origin of monetary inequality is really statistical," he told me. "It's really probabilistic. It has nothing to do with individual achievements."

Yakovenko is an affable, accomplished college professor who loves to talk. I mean, he really loves to talk, and he is so animated and engaging that one could easily spend an entire day discussing his work and theories over coffee or a few beers. Our conversation at his pleasant home in College Park, Maryland, lasted much longer than we had originally planned, as he did everything he could to talk my ear off, explaining the progression of his academic career and how he found himself exploring physical explanations for economic phenomena. It's a fascinating story.

Growing up in Soviet Ukraine, Yakovenko was something of a standout math whiz as a child, a talent that eventually drew him closer to the study of physics. Following his early graduate studies at the Moscow Physical-Technical Institute and later his PhD at the Landau Institute for Theoretical Physics in Moscow, Yakovenko's postdoc research next took him to the United States, right around the time Soviet-U.S. relations were thawing, thanks to the tireless efforts by Mikhail Gorbachev and his glasnost and perestroika initiatives. Gorbachev's thaw worked a bit too well—the Soviet Union famously collapsed while Yakovenko found himself on the other side of the Atlantic Ocean.

"In '91 I came as a postdoc at Rutgers University," he recalled for me. "By the end of the year the Soviet Union disintegrated. Me, my wife, and children, we came with Soviet passports in September and by the end of the year there was no Soviet Union, and so we had to pick our citizenship: Do you want to be Russian or Ukrainian, make your choice. That of course was a big shock."

This shock to Soviet scientists like himself reverberated well beyond basic citizenship questions. Soviet state support underpinned pretty much all of scientific inquiry in Yakovenko's homeland at this time, and with the lifting of the Iron Curtain and the collapse of communism in Eastern Europe, the Union of Soviet Socialist Republics went the way of history, and with it Soviet state financing of the sciences, including physics.

The incentive too great, Yakovenko and his family opted to stay in the United States. He found work as a research associate at the Department of Physics and Astronomy at Rutgers University in New Jersey. That led him next to an assignment as assistant professor of physics with the University of Maryland in College Park. There he's been working and researching ever since, obtaining tenure around 2000 and the title of full professor in 2004. Shortly before this, he began turning his attention away from theoretical physics and his interest in organic superconductors toward a heavier research focus on "econophysics" and macroeconomics via an idea that's been underpinning his interest in modern economic developments ever since his undergraduate years.

The idea is this: If certain laws of physics can be used to explain the distribution of energy in a complex physical system, could these same laws perhaps not also be used to successfully explain the distribution of money in a complex economic system as well? "Around 2000 I was getting tenure successfully doing this regular solid-state theory of super-conductivity," he explained to me. "That's when I started thinking about this again. It was always in the back burner of my mind."

Through his long career Yakovenko has successfully published dozens of papers, but he says his most cited, referenced, and quoted work by far is a paper he managed to get published in the *European Physical Journal* in 2000. This paper is titled "Statistical Mechanics of Money." No American publication would touch it at the time, he recalled for me.

The heart of this paper's argument is that the global distribution of money is following precisely along the lines of natural system entropy, or along a declining exponential function seen in nature when systems are left to their own devices, sans control or active intervention and input. This function, otherwise known as the Boltzmann-Gibbs law,[13] describes how energy is randomly distributed throughout the universe and explains why objects or units of equivalent qualities may display equivalent properties at lower quantities but begin to demonstrate vastly different characteristics and qualities as a whole in very large quantities. As it goes with physics, so it goes with economics and the billions of individual agents interacting in the global economy, it's argued. Yakovenko and his coauthor for this particular paper call it "the economics analog of molecular dynamics simulations in physics."[14]

It all sounds terribly complicated at first glance, but if you dig deeper you realize this concept is actually pretty simple at its core. Yakovenko's breakthrough paper argues that standard economic theories fail to anticipate grossly unequal distributions of money because economists develop and favor models that seek out system equilibrium. Such models work fine when the number of agents or units in a model is small or very limited, but they completely fall to pieces when the number of active agents or units is very large or unlimited, and this is the very reality of the modern world economy that you and I and everyone else lives in today. Yakovenko explains it perhaps best using a simple analogy.

Imagine the air molecules in the room you are sitting in right now, surrounding and engulfing you as they crash about one another and against you and the other objects in the room. Now imagine that each molecule holds identical properties, the same number and type of atoms, and they are all indistinguishable from one another (just for this analogy). When your heater or air conditioner releases these air molecules into the room, these machines are imparting energy to them, more or less "ordering" them to be at a certain temperature. Let us assume that the heater or air conditioner successfully orders these molecules into your room at an equivalent temperature, for example, 22°C or about 70°F. Now capture just a small number of these air molecules, maybe just a dozen or two of them in a container of some sort, and then observe the changes. With very small numbers of interacting molecules excited to that same temperature, you may find that these identical molecules maintain their identical properties over the course of their free interactions (crashing into one another) and during your observation of them.

This is how many economists see the world, with their models showing them equilibrium distributions much like their precious supply and demand curves or the famous model of comparative advantage pioneered by the classical British economist David Ricardo and frequently used by professional economists, university economics professors, and much of the media to argue in favor of free and unfettered trade flows.[15] These models rely on smaller-scale interactions, and the economists utilizing these models extrapolate their results to apply to the broader complicated world, assuming that, all things being equal, what works at smaller scales should work exactly the same way at larger scales as well. Piketty's simple $r > g$ equation fits this definition; it seeks to demonstrate how a small-scale model of relative rates of return best describes how the much larger world economy evolves into a matching equilibrium state, one wherein income inequality gradually widens over time.

Except the room you are sitting in now contains not just ten or twenty molecules but rather billions of them. At these much higher quantities of even identical, indistinguishable units or actors, there is no equilibrium distribution; rather, the system distribution that evolves over time is random and probabilistic. Meaning, the real world doesn't actually operate the way many common economic models assume it does. "In the neoclassical thinking, equal agents with equal initial endowments should forever stay equal, which contradicts everyday experience," as Yakovenko criticized in a 2012 paper of his.[16]

For example, Ricardo's famous comparative advantage model imagines two countries trading with one another: Portugal and England. The problem is the real world doesn't contain only these two countries; it contains something like two hundred of them. With these much larger and more complex numbers of agents, the systemic equilibrium sought out by the popular comparative advantage model and other economic modeling fails to materialize. As a result, models such as Ricardo's are blind to some of the real-world outcomes of economic interactions, such as the well-documented negative effects and unanticipated outcomes of free trade that exist.[17]

Similarly, as Yakovenko explains, smaller numbers of identical molecules will generally maintain their identical properties over time. But unlike your standard economists, physicists like him have long understood that the properties of systems begin to look vastly different at much higher numbers of individual units (or molecules, or atoms, or agents, or what have you), whether they are indistinguishable from one another or not. This greater system complexity leads to a sort of phase transition in that very system. For example, at much higher numbers the total energy among the air molecules in the room you are sitting in is no longer shared evenly; rather, it is shared along a random, probabilistic path that follows nature's tendency toward maximum entropy.

Models like Ricardo's and other economists' are "not scalable," Yakovenko explained for me. "If objects are the same you think that they should have the same energy, and that is the wrong answer." Once an air conditioner or heater releases the air molecules into a room, these molecules are no longer ordered to a certain temperature; rather they are free to interact and to distribute their temperatures and energy states in whatever manner they like. In reality they do not all maintain 22°C or 70°F temperatures and energy states, even though the room as a whole remains at this comfortable temperature. Instead, millions and billions of air molecules randomly order themselves into a highly unequal distribution of energy states or temperatures (temperature is just one measure of energy since temperature derives from the friction generated by atoms and molecules in motion). Some of these molecules will see their energy states increase to temperatures much higher than 70°F, while most of the molecules will "cool" to well below 70°F, sometimes far below this temperature. Combined, the room temperature remains at 70°F, thanks to these crashing air molecules, but if you could measure the temperature of each individual molecule, you would find some to be much hotter, while most would be colder. The ordering that evolves over time is sometimes described as a "rule of thirds," with one-third of the air molecules generating two-thirds of the energy needed to achieve a 70°F room temperature state. An observer would find the remaining two-thirds of these air molecules to be only contributing one-third of the energy required to achieve that room temperature. They order themselves this way because probability dictates that this is how large numbers of molecules will eventually order—or rather disorder—themselves over time, whether it's in the room you're sitting in now or throughout the entire universe. "The probability to have very high energy is very low," as Yakovenko explained it to me. "What is most probable is to have low energy, and that's kind of reasonable."

His 2000 paper "Statistical Mechanics of Money" argues that the world's economy works in the same manner. In other words, inequality is rising in the global economy because of this natural law of entropy.

Entropy explains why the molecules in that room organize themselves into an uneven, unequal distribution of energy states and temperatures. Entropy doesn't mean decay or disintegration, as some mistakenly believe. Rather, the process of entropy describes how systems evolve in the universe over time. Because in this universe there are many more ways to be disordered than there are to be ordered, without regular inputs of energy from an external force (like a heater), systems will naturally evolve toward a state of maximum entropy, or maximum disorder. The path toward maximum entropy in any system is described mathematically by the Boltzmann-Gibbs distribution curve, which is essentially an exponential function that works in reverse, declining rather than increasing. Imagine a line that plots the exponential increase of a population (two becomes four, four becomes eight, eight becomes sixteen, sixteen becomes thirty-two, etc.) and flip it to run in the other direction. The result is close to the Boltzmann-Gibbs distribution, or the path toward maximum entropy.

When air molecules are allowed to freely interact with one another without any outside intervention, they naturally "disorder" themselves precisely along this curve, resulting in an unequal distribution of energy states that sees one-third of the molecules commanding two-thirds of the total energy in that system, with the majority two-thirds of atoms left to share in the remaining one-third of the total energy distribution.

Yakovenko and a growing number of his followers argue that the global economy is currently behaving in exactly the same manner. In fact, investigations undertaken by him and described in later papers find that the distribution of money in the world economy very closely matches the Boltzmann-Gibbs distribution curve used to describe increasing entropy in any physical system, like molecules in a room. They find that money acts analogously to the energy in that air molecule example. Thus, in the world that you and I live in today, one-third of the world's population commands two-thirds of the world's money. The majority two-thirds of the population is left to share the remaining one-third of global wealth. The distribution of money in the global economy doesn't match the Boltzmann-Gibbs distribution curve perfectly, but Yakovenko found that over time it is gradually moving closer and closer to this line as the world economy moves closer and closer to a state of perfect entropy, or maximum disorder, a state otherwise known as globalization. He strongly believes that eventually the distribution of money in the world economy will come to match the Boltzmann-Gibbs curve almost precisely as globalization marches forward, meaning that over time, as the worldwide economy becomes "better mixed," one will find that the unequal distribution of money will rise to match the path toward perfect system entropy.

This distribution, this further "disordering" of the global economy, will mean rising economic inequality. And the distribution of money in the world economy will organize itself this way, toward a further unequal scattering, because probability allows for it, because the odds favor this very "rule of thirds" outcome.

"[The] statistical approach argues that the state of equality is fundamentally unstable, because it has very low entropy," Yakovenko explained in that 2012 paper. "The law of probabilities leads to the exponential distribution, which is highly unequal, but stable, because it maximizes the entropy."[18]

It's an ironic way to put it: freed of constraints on capital and labor, the global economy is seeking out system stability, which means greater income inequality, even though this outcome is seen resulting in much political instability lately, according to many pundits.

As money is allowed to freely cross borders and workers are encouraged to freely compete with one another both within and across these same borders, the distribution of money is matching that random distribution of energy in the molecular example. Free trade proponents fiercely advocate for releasing constraints on capital and labor, arguing that we must let the chips fall where they may and allow the market to determine the outcome. And so it does: probability dictates that the outcome that will result from this "freeing" of economic agents is a highly unequal one. Per his own estimates Yakovenko calculates that a probable, entropic distribution of the world's money supply means that just 3 percent of the world's population will find itself members of the elite, wealthy upper classes, while the rest of the 97 percent will find themselves existing somewhere along a long spectrum of exponentially declining wealth. The vast majority of people will exist at the lower end of the curve, because having less money is more probable than having more of it. The probability distribution function applies to money circulating in a freely interacting economic system just as it applies to energy circulating in a freely interacting physical one, like gas molecules in a room, Yakovenko finds.

"The probability distribution of money in that economic system should be an exponentially decreasing function of money," he recalled for me of his earlier research. "You have a bunch of agents and the probability to have a very high money balance, the probability of being very rich, is very small. There are few rich people—by definition there are few rich people. But there are many poor people." "In other words," he continued, "the probability to have a very small money balance is high and the probability to have a high money balance is low. That qualitatively looks like what it is in society: few rich people and many poor people."

To work, the Boltzmann-Gibbs distribution (or the entropy equation) requires "lots of objects and a conserved variable," as Yakovenko explained for me. Atoms, molecules, and the energy they exchange with one another meet this requirement—there are lots and lots of molecules and atoms in any physical system, and the law of conservation of energy dictates that energy cannot be created or destroyed, only changed and transferred. When an excited molecule crashes against another, it imparts its energy to that other molecule, exciting the partner molecule in this exchange. Thus the total energy is conserved: one molecule gains energy while the other molecule loses it.

The circulation of money in the global economy works pretty much the same way, Yakovenko argues. Sure, governments can print money and thus "create" it every now and then, but you and I certainly cannot, or at least we are forbidden from doing so; therefore, for a time, the money in circulation is fixed. And just like energy, this money is conserved: I transfer a dollar to you to buy your widget, and you accept this dollar and give me a widget in exchange. The distribution of this money changes— you gain a dollar, while I lose one—but the total amount of money in this simplified example remains the same: one dollar. Yakovenko calls it the "local conservation law of money."

Interestingly, separate research conducted by Yakovenko and his colleague Anand Banerjee find that this random "disordering" toward greater systemic inequality probably applies as universally to human interactions as it does to the laws of thermodynamics. Take energy consumption, for instance. There, they discovered that the distribution of energy consumption also follows that general "rule of thirds." This means that, apparently, one-third of the world's population currently consumes about two-thirds of its energy, and the majority two-thirds of the population find themselves having to split the remaining one-third or so of global energy supplies to meet their needs. It's a bit confusing because the world's energy supply is undoubtedly growing, but so is the number of people using it. As energy supply and energy demand work to even one another out, the distribution of this comparatively "fixed" energy supply is also gradually organizing itself along the probability distribution function, approaching greater maximum entropy over time, Yakovenko finds. "The exponential function does not fit the data perfectly, but it captures the main features reasonably well, given the crudeness of the data," he says in a separate paper, explaining his attempts to match a predictive energy consumption distribution model with data on national energy consumption compiled by the World Resources Institute. "The agreement is remarkable."[19] The probability distribution function also meshes nicely with carbon dioxide emissions per capita, this paper finds, which is not all that surprising, given that the world produces most of its primary energy from fossil fuels.

So why is income and wealth inequality rising? Because, Yakovenko argues, the world economy is becoming "better mixed," thanks to the modern relentless march toward globalization. As globalization's proponents work hard to free capital of its prior constraints, wealth in the world economy is scattering itself along a random, probabilistic function that naturally evolves toward a more unequal distribution. Just as entropy dictates that systems evolve from high energy states to low energy states, or from greater order to greater disorder, the world economy is following the same process, moving from greater order and equality toward greater disorder and inequality. It's doing so because that is the most probable outcome, and world economic leaders are encouraging the realization of this most likely of outcomes by encouraging greater economic openness. During this process, some countries are sliding up the curve of income and wealth per capita, China serving as one great example and other nations formerly higher up the income curve sliding down it. (This is happening to the United States today, arguably once the world's freest, most open economy and the leader of the globalization movement.) Having it any other way requires a level of government intervention in economies that no one has the stomach for anymore.

Yakovenko offers his former home, the Soviet Union, as a perfect example of this. "In the Soviet Union, as you would imagine, there was less inequality. Why? Because the economy was constrained, because it was a managed economy, it was not a free market economy," he said. "And so the result of this was that, yes, there was lower inequality because of the constraints on everything. It was all regulated." The Soviet economy was forcibly constrained by its leaders in a herculean effort toward maximizing economic equality. Now that government leaders nearly everywhere are determined to move us all in the other direction, the natural result will be a maximization of economic inequality, because probability and the laws of physics dictate that this is precisely how all complex systems will always evolve when they are allowed to do so under their own volition.

Entropy is at the core of econophysicists' thinking and their explanation for that decline of the middle class in wealthier nations that the OECD is so worried about. Globalization's fans see freer trade, freer capital flows, and more fluid migration as necessary ingredients in raising incomes and standards of living in much of the developing world, and they are likely correct. However, they never considered the possibility that achieving this would in turn necessitate a lowering of incomes and standards of living in much of the developed world. Yakovenko and his econophysics teach that this outcome is inevitable: encouraged to freely interact, agents existing in previously low energy (money) states will gain some more of it, while those previously finding themselves in higher energy (money) states will lose some of theirs, except for the most energetic (richest) 3 percent, which appear to be largely immune to this march toward maximum entropy.

Yakovenko and his students have mapped this move toward greater economic entropy for sixty countries, plotting income distribution curves for all of them. They find that these curves have been gradually shifting since 1980. Nations starting with low levels of income per capita, like China, have seen their dots on the line rise higher along this distribution curve over time, while America's dot has slid slightly lower down the line compared to where it sat in earlier years. Over time this international per capita income distribution curve is inching closer and closer to the probability distribution function curve, a line that describes a theoretical maximum system entropy.

Why is economic inequality on the rise? Because of entropy, says Yakovenko and other contrarian academics and researchers. Because, left to its own devices, the global economy has a greater probability of ordering itself toward higher levels of inequality than any other possible outcome. And so it is.

Mainstream economists and the leaders of the Bretton Woods institutions (the World Bank and the International Monetary Fund) steadfastly refuse to see the world this way, and they will continue to do so. This is because the picture that Yakovenko is painting for us all strongly implies that the policies and practices most cherished by these economic masters of the universe are the very drivers of the rising inequality and global economic stagnation that they are now desperately trying to combat or even reverse. They are seeking greater economic openness, freer trade, freer capital flows, fewer restrictions on migration, and smoother labor mobility. In all probability these policies will continue to deliver higher economic inequality and all that comes with it. After all, their advocacy for more globalization, not less, is a recipe for greater natural systemic stability, and in nature that means an evolution toward maximum entropy, which is just another term for system stagnation. Stagnant systems are fairly stable, after all.

The economic masters of the universe will refuse to believe this. Take Christine Lagarde, former head of the IMF and current president of the European Central Bank. In a September 2016 speech at Northwestern University (two months before the November U.S. presidential election that year), Lagarde acknowledged that the global economic recovery from the 2008 financial crisis was weak and uninspiring. The status quo was failing far too many people, she admitted. And yet, in that same speech Lagarde pressed for a continuation of this very status quo over any alternative approach. We all must double down on globalization and free trade, she insisted, arguing that there was no turning back and that any attempts to do so would only lead to greater economic pain and suffering for the OECD's middle-class citizens.

"Rising economic inequality is a phenomenon in many countries today, rich and poor, but it has really hit home in the advanced world right now, where real incomes for many have been declining," she said at the time. "The task at hand is, first of all, to take the right macroeconomic policy decisions and maintain economic openness, a combination that has delivered so much good for the world in recent decades. Getting everyone a bigger piece of the pie means the pie has to continue to grow."[20]

"Protectionism" must be avoided at all costs, she warned darkly. Instead, economic growth must be made more "inclusive," she intoned, vaguely. The solutions she prescribes include greater education for workers laid off by offshoring and import competition, more aid to the poor, more government debt to support infrastructure and public works projects aimed at stimulating demand, and, of course, more free trade, not less. She calls for an even deeper mixing of the world economy, with governments organizing education and retraining initiatives aimed at helping globalization's losers cope with the upheaval, while proposing some government debt-driven stimulus spending to counteract the gradual economic stagnation seen nearly everywhere. Ultimately all that's necessary is some tweaking around the edges of the status quo, Lagarde basically argued, not a fundamental realignment of governments' trade and economic liberalization priorities, policies that she and the IMF have long advocated.

"Well-designed public investment in education not only raises underlying growth but increases human capital and the earning potential of low-income people," Lagarde continued in her speech. "Here in the U.S. we have advocated raising the minimum wage and extending the earned income tax credit." There are no "silver bullets," she admits, "but if we want to keep globalization alive for the next generation there is no alternative to ensuring that it works to the benefit of all."

Further trade liberalization and further relaxing of controls on cross-border capital flows will only lead to further inequality, Yakovenko insists. And he argues that higher educational attainment for the broader population will achieve nothing. He believes that pushing workers toward better training and higher degrees will only serve to raise the minimum standard expected of worker educational attainment. The floor will move a bit higher, but so will the ceiling, he predicts, meaning a bachelor's degree becomes the new high school diploma, leaving employers demanding ever higher credentials, such as a master's degree or more complex, expensive certifications, without necessarily raising wages as basic degrees become more commonplace throughout that economy's workforce. Meanwhile, the distribution of money will continue to flow closer to a declining exponential function, he stresses, with wealth from the developed world's middle classes moving into the hands of the rising middle class in the developing world. The richest in both will remain fairly untouched by this process, and the gap between the top 3 percent and the rest will widen, Yakovenko foresees.

If there is any silver lining at all, it is this: Yakovenko believes that the world is relentlessly inching closer and closer to this perfect entropy curve, and that means it is closing in year by year on maximum economic inequality. He doubts the globe's per capita income distribution curve will come to perfectly match it—occasional government interventions, recessions, and financial bubbles can be expected to shake things up a bit moving forward. But as the world's distribution of its money supply approaches this line, growth in inequality will slow, and it may even stop widening entirely as that peak is nearly reached, he argues. This sort of systemic, natural inequality will never be reversed or corrected in any way, he believes, but it will likely cease to worsen, possibly rendering this new, more unequal reality more politically palatable to the masses most affected by it. In other words, the developed world's middle class will continue shrinking, but at a rate so slow that the trend falls out of popular political conversations. We may simply get used to it.

But, of course, every silver lining has a dark cloud.

One of the more pessimistic implications of this nature-based model of economic inequality—naturally randomized disordering of monetary exchange and wealth accumulation matching entropy—is that economic inequality as seen at present is here to stay. About one-third of humanity will continue to control two-thirds of its wealth indefinitely, leaving the remaining third of wealth and money to be split up among the majority two-thirds of the world's population. This model also implies that there will always be richer countries and poorer countries: for every Monaco there will be a Haiti and for every Japan a Cambodia.

Actually, that's not entirely correct; the proportion of poorer to richer countries will not become 1:1. Rather, this entropy-based wealth distribution model implies that there will always be more developing countries than there are developed countries (two-thirds poorer to one-third wealthier) and that the number of the world's poor will always vastly outnumber its rich so long as world leaders permit and even encourage greater economic openness, more streamlined trade liberalization, and ultimately this randomized, probabilistic distribution of wealth and income as the world's workers are pitted against one another to compete for a greater share of Lagarde's pie.

"There is very, very much a kind of profound philosophical consequence," Yakovenko mused toward the end of our lengthy discussion. "Because you see, if you just follow this logic, the probability distribution of money, the decreasing exponential function, it tells us that the interpretation now is different. It means that there are few rich people and many poor people, and it was always like this in pretty much any society." And if the evolution of the world's economy really is following natural laws long known to physicists, that means gross economic inequality will unfortunately always be with us, he added. "This simple argument from statistical physics gives us a prediction for the distribution of money, but it also means that it gives us predictions about inequality."

* * *

There have been rebellions against standard lines of economics research and inquiry before. A movement begun in France once called itself PAECON, an acronym for "post-autistic economics." The name was meant to indicate its followers' disdain for an excessive reliance on econometric modeling and complex mathematics in pursuit of imaginary worlds that had little to do with everyday experience, practices commonly pursued in the halls of university economics departments everywhere. The movement has since changed its name to "real-world economics" out of appropriate sensitivity to people with autism, and it regularly publishes the *Real-World Economics Review*, a journal dedicated to showcasing ideas that can't otherwise get published in the mainstream peer-reviewed economics journals. The most current issue as of this writing takes on modern monetary theory. Past authors have criticized the fetish for economic growth, free trade assumptions, and globalization. Other issues of this rogue economics journal have focused fire on the myriad ways in which economists ignore ecology and the enormous environmental damage ensuing from regular economic activities.

But these are economists taking on their better-published peers in an attempt to pull their discipline closer to an average person's experience and needs. What Yakovenko is doing is fundamentally different. He is an outsider scientist, a theoretical physicist no less, attempting to draw the economics discipline closer to the natural world and to the fundamental laws of physics and nature. His is a compelling argument, one that he can back up with solid evidence and strong data. Yet the mainstream of the economics profession continues to ignore Yakovenko's research and its implications, just as it does the real-world economics movement.

Thomas Piketty doesn't mention Yakovenko and his colleagues' findings or theories anywhere in his lengthy book on inequality. Their compelling ideas are nowhere to be found in semiregular World Inequality Reports; nor are they discussed in the 2014 book on "secular stagnation" published by the Centre for Economic Policy Research in London. I've never come across their mention in any report by the United Nations Department of Economic and Social Affairs, a unit devoted to exploring the economic problems of the developing world. Yakovenko admits that even the economics department at his own employer, the University of Maryland, shuns his ideas and won't invite him to participate in its seminars or public lecture circuit. His theories and models exist only in the dungeon beneath the ivory tower, quietly passed between like-minded thinkers and only allowed a brief moment in the sunlight via some obscure physics and econophysics journals before being banished below again by professional economists who insist that they know more about these subjects than the naturalists and hard scientists do.

It's ironic, really.

As mentioned earlier, economists are often accused of suffering from "physics envy," whereby they are criticized for their overreliance on overly complex mathematical models, creating worlds in their heads that don't exist in reality as they try desperately to see their discipline operate more along the lines of a natural science rather than a social one. And here we have a bona fide physicist presenting these economists with a mathematical model that very much does apply to the real universe, one that strongly matches our very own economic reality and experience, and yet they ignore him, at best, or pretend that his ideas simply don't exist, at worst. Many of these economists perhaps suffer from physics envy, but at present they are unfortunately shunning a real-life professional physicist who aspires to be more like them. I think they should be embracing Yakovenko's thinking instead, especially if these learned individuals truly aspire to understand the reality that we exist in as closely as possible. What better point of departure on this academic journey of discovery than the rules of nature, the foundational laws of the natural world?

Victor Yakovenko simply shrugs this off and continues to press ahead, happy to satisfy his own intellectual curiosity with no expectation that he's about to break through to the mainstream of the more popular economics research journals anytime soon. For now he appears comfortable and perfectly content in the dungeon and happy to let Piketty and others attract greater fame and notoriety while he goes on tinkering away quietly in his College Park office.

* * *

There are compelling reasons why insights from the natural world offer far more powerful explanations for macroeconomic phenomena than the more popular socioeconomic reasoning that dominates this discussion. That natural laws can better illustrate global economic trend lines fits into my underlying argument that the human economy is our species' ecology and is better described as the economy/ecology. So far we've explored three examples of this.

First, density dependence explains plummeting fertility and birth rates better than any other economic factor pointed to in the popular press. It underpins them all. Crowding in urban spaces exposes people to higher housing costs and an overall higher cost of living. Workers living in these conditions pursue greater levels of education and training to better compete in these crowded, hypercompetitive environments where higher salaries are necessary to achieve a more comfortable standard of living, given the higher housing costs. Longer periods spent in education and training and in establishing their careers mean couples give birth to children later, ultimately limiting the total number of offspring they produce over their lifetimes. The stress of high population densities impacts fertility via changes in both behavior and biology. Stressed individuals deliberately put off childbearing until they are more confident of their present and future economic security. They may even choose to forgo child rearing altogether, and these changes in behavior gain broader social acceptance than they did in past periods because they are a lot more commonplace now and perfectly understandable to other people finding themselves in similar or nearly identical circumstances. And the stress of life in crowded cities appears to be directly impacting biological fertility as well, likely lowering male sperm counts, for example. It's the same in nature. These effects have been witnessed in the natural world and within the animal kingdom time and time again, and wildlife biologists have long understood that density-dependent factors will impact population growth and fertility rates for a host of organisms, including insects, mammals, reptiles, and even plants. It's time to add humans to that long list of organisms that see their fertility and population growth rates declining at higher population densities. The data matches the theory perfectly as well: studies I cited earlier show that high human population density correlates tightly with lower birth numbers and lower fertility rates, more so than any other related factor. The higher the population density, the lower the number of births per woman. Mystery solved.

Though we've cracked the riddle behind our human society's plummeting birthrates we shouldn't celebrate. That's because density dependence presses down on animal population size and population growth from both ends, as I earlier noted. Thus, if higher population densities are lowering average human fertility, then I predict that they will eventually lower average human life expectancy as well. It's a dark prediction, one that I truly hope doesn't come true.

Second, falling average energy return on investment, or EROI, likely explains the steady creep of economic stagnation, or so-called secular stagnation, monitored and recorded by economists and business reporters throughout the world. The case made by biophysical economists for the dire consequences of steadily falling EROI in the oil patch is fairly strong—outside Saudi Arabia, any oil executive will admit that it takes much more effort to pull a barrel of crude out of the ground today than it did just a few decades ago. Assertions that we will soon begin witnessing consequential plummeting EROI for natural gas and coal as well are less convincing, however. Natural gas and coal reserves remain abundant in most jurisdictions, though in the United States natural gas must now be extracted primarily via the intensive and expensive process of horizontal drilling and hydraulic fracturing. Still, it makes perfect sense that broader economic activity is greatly impacted when a large chunk of the economy is devoted to just capturing, storing, transporting, and consuming the primary energy resources needed to run everything else in that economy. The net return to the economy from energy gained is declining; thus GDP growth must be declining with it, and indeed that appears to be what we are observing. At the very least the EROI model makes a critical contribution toward enhancing our basic economic understanding by placing energy front and center in the analyses. The biophysical economists are right: the economy isn't just about a bunch of actors buying and selling things from each other, fulfilling demand and supply in a never-ending cycle of eternal, unlimited growth seemingly to infinity, which is the neoclassical view of economics. Instead, in the real world the human economy is a process whereby humans capture and release energy in order to perform work, work that often entails transforming resources into substances of value to human societies, goods that are then disseminated to other humans via even more energy consumption. The economy is our species' ecology, no question about it, and all of this occurs in a finite system: Earth, with us kept alive on the surface only by the constant bombardment of energy from the sun. Energy and mittens and earmuffs and eggs are not all the same basic things, because without energy there can be no mittens or earmuffs or eggs in the first place. Or anything else for that matter. Energy is not just important. It is critical. Fundamental. The single-most important element to the functioning of our entire world economy. Mainstream, politically influential economists ignore this central fact to their peril. Well, actually, to our peril. They of course never suffer any consequences for their failures; only the rest of us do.

Now we have a third example of how natural laws that regularly apply to animal ecosystems and other facets of the natural world apply to our economic world in turn. Why is economic inequality creeping ever higher? It's called entropy. Freed from the shackles of tariffs, foreign transaction limits, and other past restraints like investment restrictions on foreigners, money can now flow smoothly across the planet with an ease never before seen in human history, often within seconds. The world's economic leaders have been continually pressing for fewer constraints on capital and freer exchange of goods, services, and people. Left to organize itself, the global distribution of money is now following a predictable pattern of a declining exponential function—the entropy equation—common to all complex systems that are left to order themselves largely without outside intervention. The path spontaneously taken is a probability distribution. In the case of the transfer of energy between molecules in the universe, the energy state distribution that results is a highly unequal one because molecules in these systems have a higher probability to exist at lower energy states than at higher energy states. Likewise, in our global economy, having a lower money balance is more probable than having a higher money balance. Thus, the odds favor greater inequality with greater economic openness, and higher inequality ensues as a result of rapid globalization. The world's economic managers have opted to not micromanage the world economy or its money supply anymore, to allow the market to sort these out on its own in whatever manner it may do so, with only occasional intervention to avert or contain crises of the sort witnessed in 2008. So the global economy orders itself naturally along a path of entropy and toward the most probable outcome: higher economic inequality. The scattering of money across the globe is not managed or planned in any particular way, so the distribution of money that emerges follows a random, probabilistic pattern seen regularly in the natural world. Entropy means change from order to disorder, and the result is a highly unequal yet more stable outcome. A condition of greater income and wealth equality is less stable and requires much more state intervention to achieve and maintain, as witnessed in the Soviet system that ultimately collapsed.

So nature holds the clues to comprehending common macroeconomic phenomena that our professional economists are now struggling to explain to the rest of us or to even understand for themselves in the first place. It does us well to listen to the natural scientists, especially the contrarian ones. Their insights can illuminate for us new possibilities, cracking wide open the answers to mysteries that have long eluded us.

As with deer and osprey and fruit flies, humans also feel the pinch of stress induced by higher population density and its tendency to depress birthrates; inadequate paid family-leave privileges, weak minimum wage laws, and even global warming have almost nothing to do with it. A squirrel struggling to acquire more energy than she exerts every day in the pursuit of food and shelter will weaken and slow; so it is with economies finding themselves struggling to acquire the new energy needed to run them. Molecules left to freely interact in a space will immediately follow a path toward a maximally unequal state of energy distribution, or maximum entropy; money permitted to freely flow across the borders of the entire world will behave no differently.

Where else can we draw further lessons from the rules of nature and how they dictate human economics and human social processes? I believe it beneficial that we focus our attention next on the other mass-colony-type species, organisms that, like us, build cities and roads, use landfills, enforce divisions of labor, and exploit resources far from their home turfs.

Ants and termites are the only other species that I can think of that build and maintain immense complex settlements of thousands, hundreds of thousands, and even millions of individuals with great success. They've been around for far longer than we have, so they have additional experience in these matters to boot. And besides, we have a lot in common with the ants, if you stop to think about it.

Engineers, workers, soldiers, caregivers, and executive leaders have all emerged in ant societies, as they have in our own. Ants are farmers and ranchers, scavengers and hunters. They dote on their young and will also attend to the injured and dying during battles with their enemies. They swarm in the aftermath of crises and set to rebuilding damaged infrastructure once the immediate threat has passed. Our cities have a lot in common with theirs, except that they mainly build theirs underground while we mainly build our urban concentrations aboveground. But we churn the earth and rearrange it into settlements better suited for our needs just as they do. Seen from space, our settlements don't look all that different from theirs: central hubs of bustling activity connecting by long paths leading to resources and other settlements in a sort of hub-and-spoke network. These patterns emerged spontaneously among the ants and termites, without any vote, conscious deliberation, or prior calculation taken by any of them. Who's to say we are any different, honestly?

Ant populations are density dependent as well. They have not yet worked out a means of escape from the iron law of EROI. And ant colonies are most definitely highly unequal: a single queen and small number of her immediate elite subjects surrounded by an immense herd of expendable workers.

Are we all just living in one giant ant farm? Let's find out.

4

THE ANT FARM

Lessons from the Other Urbanites

Throughout the world lives a common species of ant, one not yet facing any immediate threat of extinction or other serious endangerment but still of rising concern for committed scientists and conservationists. The International Union for Conservation of Nature, which maintains its famous Red List of the world's most threatened and endangered species, places the red wood ant in its "near threatened" category—worth keeping a concerned eye on but in no immediate need of intervention or conservation measures. Presumably this status holds as well for a subspecies, the Japanese red wood ant, its range running throughout Northeast Asia from the Russian Far East to as far south as Taiwan.

Unremarkable in general outward appearance, the Japanese red wood ant has much in common with other species of ants. It also has something uniquely in common with us humans as well, as I'm about to explain.

In the early 1970s Japanese entomologists made a remarkable discovery not too far from where I live on the northern island of Hokkaido, in a region famous for its long winters and heavy snowfall. On this island's shoreline roughly forty miles north of the city of Sapporo, researchers from Hokkaido University stumbled upon what was at the time believed to be the largest single ant colony anywhere on Earth, built entirely by a massive and interconnected complex society of Japanese red wood ants. "Supercolony" is a term the media used to describe it, and for a very good reason. This wasn't just an ant city; rather, it would be more aptly de-

scribed as an entire ant nation: the scientific team investigating this particular red wood ant supercolony counted more than forty-five thousand
separate nests all connected by an elaborate network of underground
tunnels. The entire colony extended along the Sea of Japan coastline for
some twelve miles. The Hokkaido University research team ultimately
determined that it contained more than 307 million individual Japanese
red wood ants.[1] That means that at the time it was discovered, this ant
supercolony's population was even greater than that of the United States.
Later development of some of the coastline in the same area is believed to
have wiped out a big chunk of this red wood ant nation, but had it
survived to half its size today, this Ishikari Japanese red wood ant supercolony would have a greater population than either Russia or Mexico.

Why would any single species, let alone tiny ants, choose to concentrate themselves in such massive numbers in just one small corner of an
otherwise very large island? Actually, the same could be asked of the
citizens of Great Britain. Despite having a whole massive island of their
own on which to stretch their legs, some fourteen million UK residents
have chosen to crowd themselves into greater London. That's about 20
percent of the nation's total population—all of those millions of people
deliberately cramming themselves into just one small corner of a massive
island. Neighboring Ireland's geographic crowding is even more ridicu-
lous—almost 30 percent of that nation's entire population chooses to live
only in or right next to Dublin. What's wrong with the rest of Ireland?

This human crowding pattern only gets more extreme if you look
farther east. Just imagine for a moment how massive Australia is—more
expansive than the continental United States—then marvel at the fact that
more than 22 percent of all Australians volunteer to live in just the Sydney area alone. That's like 71 million Americans choosing to live and
work only in New York City (whose actual population is closer to 8.6
million). And consider the extreme crowding of South Korea mentioned
in the first chapter: that nation has a population greater than California's
living in an area about the size of Ohio. Actually, you literally didn't
know the half of it: almost half of all South Koreans refuse to live anywhere except within Seoul or its immediate vicinity.

Why do we humans subject ourselves to this? If we know crowding
and higher population densities make life more expensive and painful and
leads to greater levels of stress and anxiety, then why don't we all start
escaping to the calmer, quieter boonies? The countryside is so much

nicer, after all. We idealize quiet, comfortable small-town and rural life in our literature and contemporary pop culture, yet the vast majority of us shun it all in favor of the endless rat race in the urban jungle. Why?

In fact, insects (like ants) and animals (like us) choose to concentrate and crowd themselves in this unusual manner all the time. It's entirely natural. In fact, it is by far the most recognized population distribution pattern found to emerge from the random rhythms and movements of natural animal population dynamics. Wildlife scientists refer to it as "clustering" or "clumping" and can recognize the pattern again and again in a variety of species, usually emerging as a species' numbers rise and its population becomes denser. Clemson University professor Greg Yarrow confirms that this clustering is "the most common type of dispersion," emerging especially among populations of insects and wildlife that "are often very social and live in family units."[2] Yarrow goes on to explain that this population clumping effect is a result of constraints on habitat, but that's not the entire story.

On this planet on which we all reside, food and other resources are not found scattered and disbursed nice and evenly across the landscape, to be captured for consumption with little or no serious effort or concentration. Rather, they are found in patches, and it's these patches of food and other resources that encourage the animal clustering patterns to emerge in the first place. At lower population numbers, it may appear that a population distribution is more uniform, or more evenly spread out over a particular geographic area, but even uniform population disbursements hide the fact that for most animals, the resources they need for survival are still located in these scattered and separate patches—the uniform disbursal pattern simply reflects animals' movements from patch to patch as they seek out what they need to ensure their survival. As a population grows larger, its members' reliance on these food and resource patches becomes more acute given heightened competition, and so they begin to hang around the choicest patches on a more or less permanent basis to ensure their access to this supply. Thus, we see animals forming clusters or clumps of populations, the tendency to concentrate themselves into smaller geographic areas, despite having a lot more total space to play in.

As population sizes increase, clumping progresses ever forward and begins to take on a whole new logic all of its own after a while. Eventually this pattern becomes enticing in its own right for the animals involved. For example, they may come to find that it's actually an ideal strategy for guarding themselves against predators—the famous strength in numbers. It may facilitate better reproductive success, with clustering populations evolving ways to share in child rearing and child defense. And though higher population density and concentration elicits competition, as outlined in Chapter 1, clustering or clumping also enables better cooperation, especially in matters of food discovery and distribution. For the longest time it was explained that species of vultures find their life-sustaining carrion thanks to their extremely keen sense of smell and excellent vision, and that's true. But today researchers understand that most of the time, vultures find their food simply by watching other vultures. These are scavengers, after all, not hunters, and apparently their evolutionary path has led them to share rather than hoard. That's why they circle over carcasses and roadkill: it's their way of ringing the dinner bell, alerting others nearby as to where food can be found rather than keeping this secret all to themselves. This is smart, because more often than not the single vulture who happened to stumble upon dinner that day will find itself hungry and in dire straits down the road, eventually forced to rely on the generosity of its nearby fellow vultures to locate a bite to eat. If they all circle before scavenging, they all benefit, and the closer the vultures keep to one another, the better their chances of spotting a meal thanks to the discovery of one or more neighbors.

Not all animals cluster, but most of them do, even many birds of prey, animals that we normally imagine spending most of their lives in solitary settings, soaring high above the land, scanning vast expanses of brush for prey to swoop down on. A 2007 study into the clustering behavior of owls made a point to correct this misimpression. "Although most bird species nest as solitary pairs for purposes of concealment or territoriality, many birds have been observed to nest colonially in close association with each other," the authors inform us. "Nest clustering and distribution patterns may be due to a clustered food source, limited suitable nesting habitat, habitat continuity, and/or reduced predation risk due to owls alerting each other to predators."[3] The authors go on to clarify that crowding among owls also engenders competition, just as it facilitates cooperation to the benefit of greater group survival.

Animals such as large herbivores will cluster around ideal food sources and will maintain these grouping patterns for defense, better reproductive success, cooperation in the search for new food, and various other reasons. Meanwhile their predators (wolves, coyotes, etc.) will then display clustering patterns of their own, grouping together and keeping close to an agglomeration of prey animals that serve as a conveniently clustered food source. Even parasites display clustering behaviors. A 1995 paper I stumbled upon in the course of researching this book investigated clustering patterns for two species of aphids, pests that feed on medicinal goldenrod plants occasionally used to make herbal teas. The scientists in the study mention an important lesson of population dynamics that I alluded to earlier, how greater population density leads to greater vulnerability to predation or to parasites, two decimating factors that help put a check on population growth. This particular study discovered that dense concentrations of goldenrod plants were more susceptible to clumped aphid infestations compared to solitary plants. [4]

Just as the wild animals tend to cluster in ever greater numbers as their populations rise, so do we human animals. Studies tracing the ancient path of Homo sapiens out of Africa theorize that early humans first followed the coastlines and settled at or near sources of fresh water, lots of times where rivers met the sea. There is pretty good evidence that the early migrants out of our birthplace in the Rift Valley became rather hooked on consuming coastal shellfish such as clams and oysters, animals high in fat and protein that can't easily run away, and of course the fresh water would have been a critical necessity. Eventually early migrants followed the path of freshwater rivers and streams inland, leading them to eventually populate the interiors of landmasses. These early human wanderers inevitably came across ideal patches of food and then settled at or near them. As their numbers rose and rose, additional population pressures likely prompted either more out-migration or innovations with new and additional food sources, adaptations that engendered even further group success and survival.

This more likely than not explains the origin of agriculture.

I was forever taught in my grade school lessons (as perhaps you were) that agriculture came first, the towns and cities second: that the ability of humans to grow and store surplus food enabled the rise of crowded and more complex settlements. In reality, the cause and effect almost certainly worked in the exact reverse: the towns and cities grew first, which led to pressure on populations to find additional sources of food, prompting early humans to eventually experiment with growing their own. At least that's how it works in nature. I have a really hard time believing that ant species that grow their own food (yes, they do exist) evolved these behaviors before crowding themselves into anthills and not afterward. In nature the clumping comes first, and then new adaptations and evolutionary behavioral changes always follow. It doesn't work the other way around for animals, so why would it for us?

From here we can paint a fairly good picture of how and why our modern cities rose up.

You'll notice that the world's largest urban concentrations are generally found at or very near the coastlines and are adjacent to one or more sources of freshwater. Sometimes this fresh water source is underground. These human clusters also appeared and expanded adjacent to other resources to make them easier to exploit for humanity's evolving requirements, such as prime agricultural land, forests, choice rock quarries perhaps, and particularly rich fishing grounds (or a number of other resources that we later determined to be essential; the cities of Midland and Odessa in western Texas wouldn't exist were it not for the ocean of crude oil they sit on). In other words, humans located the best "patches" of resources and then built their settlements there. More settlers were attracted to these patches. Eventually this crowding took on an entirely new logic of its own as humans found these clumps of fellow humanity most favorable for reproduction and defense, among other cooperative ventures that improve our species' chances of long-term survival, such as trading goods and services. Over time trade became a critical component of clump success, enabling conveniently located coastal cities to grow into even more massive conglomerations of Homo sapiens as trade crossed the seas and oceans.

So now we have the answer to a few questions that I posed to you much earlier in this book. Why did we fan out of Africa, our initial birthplace, where the climate is so much better suited to a hairless species of ape such as ourselves? And why on earth did we eventually crowd ourselves into massive and enormously stressful urban concentrations, so stressful in fact that we've now sharply lowered our reproductive potential in an inadvertent collective bid to cope, drastically weighing down on a key driver of our species' very biotic potential? The answer, of course, is because the population clustering or clumping patterns common to animals and insects are common to us as well. The dynamics are the same.

We clustered because that's just something that animals like us do. The rules of nature once again determined humanity's fate. Urban crowding eventually took on a life of its own, until eventually we evolved our human ecology into a complex, interwoven global economy (the economy/ecology, or vice versa) with the metropolitan city at its center. This economy/ecology is driven by the goods and services we trade with one another in order to determine how resources are to be collected and redistributed throughout the world's population, with humans landing on money as a means of exchange used to determine how this resource allocation is to take place. The resulting allocation is highly unequal because, as we have learned, all pattern distributions in nature are highly unequal as systems follow a natural pathway toward maximum entropy when they are left to run freely and uninterrupted.

Energy is the very first critical component. Energy stands at the center of this global human ecological/economic strategy: we use energy to transform matter into materials and goods that we find particularly useful and of value, and these goods are then redistributed (using more energy) in accordance with which members of our colonies have the most or least money. This money we've acquired through both competition and cooperation. It takes energy to find and get new energy; thus easily acquired energy facilitates greater ecological/economic stimulus and activity. However, energy is becoming harder and harder to come by lately, at least in the case of crude oil, which is the principal form of energy running modern civilization. Therefore, our ecological/economic activity and even our rate of new colony formation are now slowing down precipitously, a phenomenon we observe as the falling rates of annual gross domestic product expansion. Meanwhile, density dependence (caused by the very crowding we all unconsciously forced ourselves into) is now pulling our rates of reproduction down to far below what they were in the past, putting an end to that exponential growth curve of human population expansion that characterized so much of the nineteenth and twentieth centuries. That curved line is now flattening. It will eventually flatten out and plateau altogether and may even bend in reverse at some point.

So what does all this have to do with ants? Everything.

The ancient ancestors of ants resembled something closer to modern wasps (in some modern ant species, their drones, the only males in a colony, actually still look like wasps). These ancient wasplike insects clustered together many millions of years ago, as expected, and eventually evolved into the vast array of ant species that we see in our modern world today—at least eleven thousand different ant species have been identified,[5] and more will be discovered. As they are subject to natural patterns of clumping, so they are subject to some other rules of nature that humans find themselves unwittingly bound to as well, rules that I've been outlining for you in this entire text. Their populations expand exponentially, but not forever. Their resulting very high rates of population density and population concentration give ants very low average fertility—the worker ants don't reproduce at all; only the queens do. They experience general ceilings on colony size given the environmental constraints they find themselves in, especially the constraint of diminishing energy return on investment. Ant societies are also apparently highly unequal, and the vast majority of their members are wholly expendable. Millions could have perished from that Ishikari ant nation at any given moment, and the survivors would have barely noticed. And yet it's possible that this expendability and redundancy is one of ants' greatest strengths. These highly organized, highly complex ant societies manage to build great cities, even entire countries, and in many cases come to utterly dominate the immediate environment that they colonize and conquer, sending other species running for cover. It's almost like a mirror image of humanity—almost.

For starters, they've been around for quite some time. In fact, ants have existed on this planet since long before the dinosaurs came on the scene, first emerging about 140 million years ago.[6] So they've been at it quite a bit longer than we have. Perhaps we have many lessons to learn from these much more experienced organisms, Earth's original urbanites. The parallels between our worlds and theirs can be quite compelling and endlessly fascinating, but let's try not to get too carried away with them.

* * *

As you may have gathered by now, one overarching theme of this book is that we humans may not be in control of our ultimate fate, mainly because we probably aren't nearly as smart as we think we are. Our societies' principal choices, which we think we've been making collectively and consciously—how and where to build our cities, how many children we give birth to on average throughout the population, the most efficient way to distribute the fruits of our modern global economy—aren't really choices at all in reality; rather, they are compelled by natural forces that apply to all animals on this planet, whether we recognize this or not. At least that's one argument worth considering.

A corollary to this may be that the ants on this planet aren't nearly as dumb as we might think. Popular impressions hold that ants, like so many other insects, are essentially organic robots, preprogrammed via natural instincts to perform specific tasks in aid of their survival and reproduction. Thanks to more thorough modern research into ant behavior, we now know better. Surprising as it may seem, scientists have observed ants demonstrating higher cognitive functions, like a capacity to learn and to memorize things. They have situational awareness. They teach each other and can demonstrate complex problem-solving skills, both individually and collectively. Some of them use tools.[7] They may even be somewhat curious about us: In her book *Ant Encounters*, entomologist Deborah Gordon tells of a colleague recounting for her how he once observed a single ant actually observing him while he was out conducting field research. The little ant apparently stood completely frozen and stared at the scientist, seemingly enthralled by him, observing this researcher's every move, while other ants simply scurried around busying themselves with their chores.

Ants can even display what some may describe as empathy for their fellow ant sisters. One bit of evidence for this apparent alternative ant intelligence could be what scientists have described as the "rescue behavior" practiced by some of them. For example, a 2017 paper investigated the lives of a species of ant fond of raiding termite nests. These termites are formidable opponents, and some of the raiders will inevitably die. Fascinatingly, these ants were found to carry their injured back to their home nests for recuperation, giving the battle-injured time to heal and fight again another day. "After a fight injured ants are carried back by their nestmates," the authors informed their readers. "These ants have usually lost an extremity or have termites clinging to them and are able to recover within the nest."[8] The authors interpret this as an evolved behavior that allows the ants to minimize the costs of their raids.

I've had my own encounter with animal empathy in the wild. During a reporting assignment in southern Kenya, a work colleague, two armed park rangers, and I set out on a mission through the bush to locate a lion mortally wounded at a Maasai village the evening before. The idea was to locate her so that the veterinarian flying in from Nairobi could get to work trying to save her life. We found the injured lion, but that's not all we found: we were unable to get too close because a second female lion sat there right next to her. She didn't appear to be guarding her; rather, she seemed more intent on comforting her and doing what she could to ease her pain and plight (that lion sadly died the next day).

So it may be with ants, in a way. In fact, the long-held image of ants as reflexive, stimulus-driven robots is quickly being shattered. Anna Dornhaus and Nigel Franks, two researchers at the University of Bristol, wrote a decade ago that contemporary studies are finding out that ants and other insects can display surprisingly complex cognitive abilities considering the extreme tininess of their brains. "This involves not only simple conditioning to the locations of stimuli associated with food, but also more complex learning, attention, planning, and possibly the use of cognitive maps," they reported.[9] Ongoing research into ant and insect intelligence holds the promise of potentially brilliant scientific breakthroughs. Dornhaus and Franks see "an opportunity to link brain structure and function, and even to study properties of individual neurons," given the small size of their nervous systems and brains and the relative simplicity of ant behavior.[10] Understanding how thought evolved in ants could be key to better understanding how a wide array of other animal behaviors has emerged via evolution and natural selection, such as the elaborate courtship dances many bird species conduct, for example, or even farming and ranching. Some ant species practice both these behaviors, as do we.

Ants can also change jobs, just like we do. For sure, biology locks some of them into certain unalterable functions, with ants divided into morphologically different queens, drones, and workers. The jobs of queens and drones are fairly fixed, but the worker ants don't all do the same thing each and every day of their short lives. Researchers note that many ants seem to stay indoors more or less permanently to take care of the queen's larvae. Others scurry back and forth collecting and transporting food to the colony. And other workers are kept busy building or repairing nests. But they can switch roles depending on the situation: the child-care givers will emerge into the sunlight on occasion to help forage, foragers will often stop what they're doing to aid in nest repairs, and members of the construction and maintenance crews will chip in every now and then to go down below, check in on the queen, and help feed and care for the colony's young.

Gordon argues that we should dispense with the notion that there is a sharp division of labor within ant colonies. Rather, her research has identified a pattern of ant behavior she calls "task allocation," which she describes as "the process that adjusts the number of ants performing each task according to the current situation, both in the world around the colony and inside the colony."[11] Gordon objects to the classical "division of labor" concept that economists are all too familiar with, an idea brought up by Adam Smith that should probably be relegated to the dustbin of history, given the lessons nature provides to us: "Division of labor, which implies specialized individuals, is only one of the many ways that a colony could accomplish task allocation, and it evokes a static procedure in which each individual is permanently assigned its place on the assembly line," Gordon explains. "I chose 'task allocation' to focus on the collective outcome of colony organization, however it is accomplished, and to emphasize that colonies shift their behavior in response to a changing world."[12]

In fact, oftentimes ants will adjust to changes in their environment better than we do. If an ant hill is damaged or destroyed by rushing water, the ants will quickly get to work rebuilding their nest, only away from where this initial damage occurred, usually putting it up on higher ground. We humans will rebuild flood-destroyed homes again and again on the very same spots they were swamped out of in the first place. You may have read fairly recent news of hundreds of homes, most worth well over $1 million each, being destroyed by lava from the latest volcanic eruption on the eastern shores of the Big Island of Hawaii. The most aggravating part of this story is that these homes were built on top of a lava field laid there by that exact same volcano in the 1960s. These mini-mansions might as well have been built on a pool of gasoline next to a sparking campfire—the risk was that obvious.

As Gordon explains in her fantastic book, ants can alter or adapt their behaviors based on what other ants are doing, and various interactions shape the overall form and function of an entire ant species or even just one specific colony. No one ant is hardwired or preprogrammed to perform specific tasks or to bend to a preassigned division of labor. Ants assume their roles based on their interactions with other ants and the cues they receive from them, just as you and I probably landed in our careers because of the various influences we've been exposed to over our life-times. A greater ant colony's behavior isn't preordained or organized in a rigid manner by some force of evolution that our brightest scientists haven't deciphered yet; rather, it appears or arises spontaneously and over time.

Gordon, who runs a lab at Stanford University, calls this concept "emergence." The idea holds that we should not interpret the behaviors of some social animals and insects, like ants, as reflecting an arbitrary top-down organizational structure or some rigid hierarchy that evolved in that species somehow. Rather, she believes that we need to accept that the broader behavioral picture we are witnessing and describing in the first place is the product of various complex interactions of the individual parts of that society and that the whole picture is an emergent outcome, resulting from a more bottom-up process. The whole is the sum of its parts, but those parts are constantly changing and interacting with one another, resulting in the whole picture that we are viewing. "The behavior of an ant also depends on its interactions with other ants," as Gordon wrote. "This doesn't mean that there is some colony soul that directs the ant's behavior. The behavior of the colony *is* the sum of the behavior of the ants, but the behavior of each ant depends on more than its own attributes."[13] By understanding this, entomologists can describe how millions of different ant interactions result in a particular emergent phenomenon: the behavior of the entire colony as a whole.

Emergence thinking is an excellent means of analyzing and describing the development and function of complex networks that lack any centralized control, such as socially complex animal networks and how they come to organize themselves and operate. On the front page of her Stanford lab's website, Gordon describes how she and her colleagues use emergence concepts derived from studies of ant colonies to investigate other decentralized complex systems "such as the internet, the immune system, and the brain."

With ants themselves, it's better to understand that the queen of any colony doesn't actually "rule" over her "subjects" in the classical view that Adam Smith and other early philosophers would recognize. In reality, she finds herself at or near the heart of a sort of decentralized, bottom-up dictatorship, one in which the workers are mostly in charge, even if each individual worker is in reality expendable and easily replaced. Their rule is unintentional, of course. This is the picture that has arisen over the course of ant colony evolution, an emergent feature of countless individual ant interactions summing up to a whole colony's behavioral patterns. In ant colonies the queen is not in charge; rather, she's a sort of elite slave or a cow, pampered and protected for sure but hardly free to pursue her own interests or ambitions or to even leave. She's confined to her lot in life by the thousands or even millions of workers who put her there, and she just sort of goes along with it.

It kind of reminds me of the Japanese royal family. They may enjoy a life of opulence and wealth, dining with the jet set and world famous, but their schedules and daily routines are tightly controlled affairs managed by the Imperial Household Agency. This agency is essentially a group of government bureaucrats who dictate practically every detail of the Japanese royal family's life, from whom they can meet and eat with to how they must present themselves in public and even whom they may marry. When the Heisei emperor Akihito wanted to abdicate the throne and give it to his son, he had to ask Japan's parliament for permission first—and they almost didn't give it to him. Some lawmakers were perfectly happy with the idea of forcing him to continue as emperor against his own wishes until he died, but thankfully common sense prevailed, and Japan is now in the Reiwa period under Emperor Naruhito. Or is it Naruhito who is under Japan?

Ant hierarchies are extreme versions of this sort of phenomenon, of course, and not directly comparable to their human counterparts. Queen Elizabeth II hasn't been trapped in a dungeon for decades, forced to have sex with a bevy of elite strange men in order to keep the royal bloodline pure, although I sometimes wonder if many in the British tabloid press wouldn't advocate for such a thing. But given that "leadership" (symbolic or no) is hardly that in many natural settings, such as in an ant's world, this should perhaps raise questions as to what it really means to "lead" in human societies as well and how we organize our particular hierarchies.

When studying political science in college, I often wondered why millions of individual humans permit themselves so easily to come under the thrall of some single charismatic candidate, politician, political party, or tyrant. We value our freedom and independence and desire a maximum of both, so much so that we enshrine them in our constitutions. So why do we so often volunteer to become the unblinking, unquestioning subjects of an elite political class? Perhaps I've had the ordering wrong all along. Perhaps I and billions of others like me are the "workers" in this analogy. Have we really been shackled by the elite, or did we shackle ourselves? And are the elite, in turn, shackled to us, finding themselves in their status only as a chance consequence of the way we have organized ourselves over the course of millions and billions of individual interactions spanning centuries until we've arrived at where we are today? Are our human hierarchies arranged and enforced structurally from on high, or have they emerged organically from below? Dictatorships shouldn't be excluded from this point of view either. In the lead-up to the disastrous U.S. invasion of Iraq in 2003, *New York Times* columnist Thomas Friedman once asked (and then promptly ignored) a very important question: Was Iraq the way it was because of Saddam Hussein, or was Hussein the way he was because of Iraq? Friedman was probably on to something and didn't realize it at the time.

A cursory look at the state of democracy versus dictatorship in our world reveals that complex higher-order economies that consume more energy per capita and command more resources tend to organize themselves along more decentralized, participatory democratic lines, where political power is periodically contested and disputes are settled in as neutral a manner as possible (most of the time). This suggests that social democracies emerge from greater economic and resource-consumption complexity. More despotic states tend to have simpler, lower-order economies with much lower rates of energy consumption per capita. The diffusion of less energy consumption per capita throughout a society perhaps requires more centralized control in order to keep that society more cohesive; thus authoritarianism emerges instead. The exceptions appear to be states like Saudi Arabia, which is extremely despotic and corrupt but can afford lavish energy consumption because it is literally sitting on the world's primary energy source; thus its access to this most valuable patch is secured.

If the above model is correct (and I'm not saying it is), then this likely spells doom for the single-party dictatorship running (or ruining) things in China at the moment.[14] As China's energy consumption per capita increases, the Chinese "colony," as it were, may simply require more decentralized control of its economy/ecology and its politics in order to facilitate a smoother functioning of things as the nation achieves higher orders of economic complexity. Either that, or progress toward that goal in China will hit a wall in the form of a spoiled and selfish elite unwilling to part with its rigid social controls and refusing to share political power, resulting in an outcome somewhat like, but not identical to, what occurred in North America in 1776.

What about work? How do complex social animals evolve the complex roles that make their societies so successful?

In her studies of task allocation, Gordon discovered that ants would adjust their work roles—forager, maintenance worker, or caregiver—depending on changes to their environment that she deliberately introduced to disrupt their daily routines to see what happened. If she littered their nests with junk, foraging effort would drop as foragers were recruited to help clean up the mess. If she disrupted foraging, more ants would be recruited to increase foraging effort to make up for the disruption. She concluded that ants make split-second decisions on what tasks to perform based on how others around them are behaving. Perhaps most fascinatingly, entomologists have discovered with some ant species that numbers of ant foragers, maintenance workers, and caregivers do not necessarily grow in proportion to the expansion of the overall colony's population size. Studies Gordon cites in her book reveal that ant colonies apparently only need a certain number or proportion of foragers to keep their queen, her brood, and the colony at large fed and happy. Forager numbers did not expand in parallel with colony population size. Rather, even with greater population numbers, the number of individual ants performing foraging, maintenance, child care, or other tasks (some patrol for defensive purposes) can stay fixed or grow much less slowly than the population at large.

So an ant colony's population can rise, but the colony will still only use a fairly fixed or slower-growing number of foragers, nest maintenance crewmembers, and other types of workers in the ant labor force. Well, what happens to the other ants left over? Apparently they drop out of the workforce entirely. They basically become unemployed.

"As an ant colony grows, if all the new ants are not foraging, what are they doing?" Gordon asks in her book. "Surprisingly, the answer seems to be that they are doing nothing."[15] She then spends the next few pages trying to rationalize this phenomenon, to find any reason or practical purpose for this rising ant unemployment issue. Why do large numbers of ants appear inactive and, as she puts it, just sort of "hang around"? Perhaps, she suggests, these workers are on standby in case of an emergency or are a means by which the colony can store energy or food in reserve. I have an even better suggestion. If emergence can see the evolution of specific ant colony social behaviors whereby ants can change jobs depending on the needs of the larger ant society, then it only stands to reason that a pattern can emerge through this same organic, bottom-up process in which they lose their jobs altogether as well. I mean, that's what happens in the human economy: jobs are created, changed, and sometimes destroyed. We humans don't bother attributing any rhyme or reason to it all, though we do strive to keep unemployment to a minimum.

However, we are also much better at developing a greater variety of career paths for the members of our species. The world's workforce boasts thousands of different job descriptions, and the more complex our economy becomes, the more diverse the workforce grows, given our ever-evolving needs and desires. To list all possible U.S. occupations alone would likely take up half the pages in this book. In fact, hundreds of job categories exist today that simply didn't when I was a child.

For ants we require much less space to describe the various jobs they are allowed to pursue in nature. As mentioned above, queens and drones are queens and drones no matter what. Things get a bit more flexible from there, but not by much. Ants are foragers, builders and maintenance workers, caregivers, and patrollers. They are hunter-gathers and scavengers. Some are still nomads. They don't make tools, but they can use them. And, as noted before, they can also be farmers and ranchers. So if you ask a young worker ant, "What do you want to be when you grow up?" you shouldn't have to wait too long for an answer—there are only so many paths for this little gal to choose from.

Arguably, humanity's greatest innovation is the ability to grow and/or raise its own food. Agriculture is what makes possible our enormous human population—some 7.7 billion strong and counting. Innovation in agriculture has also seen our species beat Thomas Malthus's dark predictions of impending mass starvation for future generations. Of course some issues should concern us. For instance, phosphorus supplies needed for making fertilizers are becoming rather tight, unfortunately, pointing to a forthcoming economic windfall for Morocco, which is home to something like three-quarters of the world's known exploitable natural phosphorus reserves. And though agribusiness concerns like to boast of their continuous leaps and bounds in technology and innovation, in reality we seem to have hit, or are close to hitting, a ceiling in terms of raising crop yields per acre or hectare. Not to mention the fact that animal husbandry of the mass-factory sort is one of the most polluting industries, responsible for huge water-quality problems, and the push to expand more areas under cultivation is eating away at precious rainforest in South America and Southeast Asia, among other locations. But it is something of a modern miracle to see piles of delicious-looking fruits, fish, grains, and meats flown in from across the world and available for purchase at fairly inexpensive prices at your average neighborhood grocery store. In my little corner of the world, northern Japan, I can buy cheese from France, wine from South Africa, avocados from Mexico, beef from Brazil, coffee from Indonesia, and soy from North America, all at the same place within the same hour. The vegetables and eggs are usually obtained closer to home due to spoilage risks, but the fruits can come from all over the world, including bananas from as far away as the Dominican Republic. Good imported beer is harder for me to come by, unfortunately.

Without agriculture we would still be foraging and hunting in nature for our nutritional requirements on a daily and even hourly basis, thus putting a hard cap on our population size. Indeed, the ability to grow and raise our own food is the very thing that probably allowed us to conquer this world in the first place. Agriculture has made us by far the most successful species on the planet.

Yet again, human experience is merely an echo of the natural world. As it turns out, the most successful ants on this planet are also farmers. In fact, you've probably already heard of them, as I am hardly the first one to tell of their remarkable exploits.

Leaf-cutter ants can be found throughout Latin America and in some parts of the southern United States. They're known for routinely building massive colonies, not necessarily on the scale of that ant nation discovered in Hokkaido but at least on the scale of an ant city-state, with a single colony easily consisting of up to eight million members (comparable to the population of Switzerland and certainly more individuals than live in Singapore). They're ubiquitous in the warmer climates of the Western Hemisphere and considered the most successful type of ant around. Indeed, they're almost certainly the second-most populous and successful single species of any other on this planet, aside from us, of course. Many consider them pests for their ability to strip a tree of its leaves in fairly short order, hence their name. But they don't eat the leaves. Instead, the leaves they cut and carry back to their nests serve as fertilizer, food for a particular type of fungus they grow in elaborate underground gardens. They then eat this fungus, a source of nutrition that's of particular importance to their queens and larvae.

Scientists figure that the ancestors of modern leaf-cutter ants made the transition from hunter-gatherers to farmers beginning around fifty to sixty million years ago,[16] much sooner than us of course, but it likely took them millions of years to evolve into full-blown farmers, whereas our kind probably managed that transition in only a couple thousand years. There are probably lots of theories as to how this all transpired, but one of them holds that the ants were first introduced to the fungus when it accidentally grew inside one of their nests, and they eventually found it to their liking. Cultivating, nurturing, and disseminating this fungus now lies at the center of this ant species' ecology, and its members take great care to protect their precious fungus gardens from any threatening pests or parasites that might try to destroy them. However they learned how to farm, the move toward growing their own food, rather than remaining restricted to foraging for it like other ants, is as central to the leaf-cutter ant's success as it is to ours, entomologists agree. The same goes for other fungus-farming insects, which includes other ant species, some termites, and a few species of beetles.

And just as the advent and rise of agriculture made our societies far more complex than they might otherwise have become, agriculture appears to have had the exact same effect on these farming insects much earlier. "The transition from a hunter-gatherer existence to an agricultural life has resulted in both the fungus-growing ants and termites evolving into extremely successful insect societies, exhibiting substantial diversity in colony organization, widespread distributions, and large colony sizes," concludes a team of researchers from the University of Kansas, University of Texas at Austin, and the Smithsonian. [17]

This means that leaf-cutter ants have a more complex social order and division of labor compared to other ant species, comprising what scientists like to describe as "castes." Though they still exhibit the task-allocation behavior identified in Gordon's research, their potential career paths are a little more narrowly defined because of the ants' polymorphism: they come in different shapes and sizes depending on what job they were more or less "born" to do. Science also regards their cultivation of a particular type of fungus as representing a symbiotic relationship. Sure, they eat the fungus, but they also care for it and spread it from nest to nest, likely making its occurrence in nature far more common than might have otherwise been the case had these ants never evolved this behavior in the first place. Some writers have characterized humanity's obsession with certain crops and plants as something akin to symbiosis as well. If species' "success" is classified as a high propagation of genetic material, then I suppose you could make the case that humans are as good to bluegrass, petunias, and asparagus as those plants are good to us. In the case of fungus-farming bugs like leaf-cutter ants (and many others), the scientists previously mentioned certainly see their care and consumption of this fungus as a win-win situation. "The fungus garden serves as the primary food resource to the ants, especially for the queen and larvae," they explain. "In exchange, the ants disperse the cultivar to new colonies, provide it with substrate for growth, protect it from pathogens, and enhance its growth."[18] Putting it that way, then I suppose one must concede that the deal is pretty good from the fungus's point of view as well.

There's more. Farming has also permitted leaf-cutter ants to evolve into the longest-lived ants on the planet. Your average ant will last about a year or so. Take good care of your pet leaf-cutter ant, and she could stick around in your household for at least as long as your cat.

Ranching ants, other early agriculturalists, have had a pretty successful run as well, though less so than ants that evolved dominant farming cultures. This too pretty much matches human experience—by and large the farmers were the ones to build and spread great empires, outpacing the pastoralists. But there are lessons for us to draw from ranching ant cultures.

Some species of ants cultivate aphids, doing their best to protect them from predators while the aphids graze on their food of choice, such as goldenrod stems. They'll even occasionally move their aphids to better pasture. Scientists in Japan recently discovered that the aphid-ranching species *Lasius japonicus* will favor one type of aphid over another, improving the former's numbers in the wild.[19] Some press reports call these aphid ranchers "insect cowboys," but that's not a great analogy. Cowboys care for cattle on their ranches because the cattle are to be slaughtered later for their meat. This isn't the ants' intention—no aphid slaughtering ensues once the aphids are sufficiently fattened up. Instead, these ants are dairy farmers: while tending to their herds, the ants will stimulate the aphids with their antennae, causing them to secrete a sweet honeydew substance. The ants then drink this substance and also carry it back to their nests, thus sort of "milking" the aphids.

Studies into the unique behavior of aphid-ranching ants are ongoing, but solid evidence is emerging that this ranching adaptation has led to greater social complexity in these ants, just as farming did for the leaf-cutter ants and other fungus-cultivating ants and insects. This could demonstrate that, in nature, more complex resource appropriation leads to animals evolving more complex social arrangements and hierarchies, adaptations that improve resource-extraction efficiency. Here we have yet another pattern seen in nature that maybe (just maybe) helps us to better understand our very own human world. If higher-order resource extraction and consumption leads to higher orders of social complexity in ants, could this not also explain the evolution of social complexity and hierarchies in Homo sapiens and the human economy, replete as it is with multiple social divisions and tasks (and incomes)? I think it's possible.

"Investigations carried out on a limited number of ants and species of aphids have shown that the behavior of foragers tending aphids (aphid milkers) differs significantly in different ants," as one paper on the subject put it. "The interactions of ants with aphids can be characterized by their different degrees of functional differentiation with the aphid milkers ranging from unspecialized to 'professional' foragers."[20] For example, in some aphid-ranching species some ants act as the "shepherds" and others as the "guards." Others still are designated as the professional "transporters," tasked with delivering the honeydew back to the nests. There's a reason for this higher order of social complexity: the greater specialization or more specialized task assignment that appears to have evolved in some aphid-ranching ants "seems to increase the efficiency with which they can collect honeydew,"[21] scientists explain. Furthermore, the degree of specialization also appears to increase as a ranching ant colony's population increases and as pressure on its existing food resources (the aphid herds) mounts, research has also discovered. The higher the colony population, the higher the number of "task groups" within the ranching culture, this research finds, in a strong positive correlation.

The paper quoted above even points out a seasonality apparent with the ant ranch hands: their work slows down in the fall, then perks up again in the summertime. This isn't terribly surprising, I suppose. An aversion to colder weather is actually found in a host of other ant species, including ants that never learned how to farm or herd aphids. Research in Virginia, for instance, discovered that common forager ants will adapt their behaviors depending on outdoor temperatures and even certain times of day. The ants studied slacked off and neglected their foraging duties almost entirely when the thermometer dropped below 10°C—that's only 50°F (which really isn't that cold, when you think about it). The paper also found that common forgers don't like it too hot either. On hotter summer days, they did most of their foraging in the cooler morning or late-afternoon hours. During the heat of the day, they all took a siesta.[22] In a similar vein, the ant ranchers may find the metabolisms of their aphid herds slowing down when the temperature drops. Either that or they simply don't like working too hard in the cold. And who could blame them, honestly?

The broader point is this: putting it all together, what lessons can we draw from ant agriculture? With ants and humans, greater social complexity not only arises from higher population numbers but especially ensues with more advanced and complex food and resource procurement practices—meaning agriculture, both farming and ranching—and as pressures on these food resources push ant species to even further specialization of their labor roles in order to maximize the efficiency of their resource extraction. In other words, there's evidence that in animals and insects more complex resource-acquisition strategies tend to result in greater complexity of social organization and social hierarchies, hence greater complexity in a species' economy/ecology. And so it is with ours.

It must be the same for energy. For ants, their food is their energy source as all ant labor is performed manually, sometimes with the aid of tools but without any assistance from machines or draft animals. Farming is more complex, so a greater variety of tasks for ants emerges even as they continue to rely chiefly on their own labor and muscle power to move this higher-order ant economy. Looking at things from this perspective, then I can begin to see why that Finnish paper I cited earlier, in which the authors advocate for a complete shift to renewable energy, argues that this will likely lead to lower income levels in developed nations but at the same time a more equitable distribution of that income. The early days of simple fossil fuel extraction—merely digging or drilling a hole and gathering the rock or goo that came out of it—led to the rise of a greater variety of tasks and employment needs in humans than what we saw with simple wood burning. As fossil fuel extraction became more difficult and complicated, the fossil fuel industries' divisions of labor, technology, and capital requirements grew more complex and onerous in tandem, all the way to deepwater and ultra-deepwater offshore drilling and onshore hydraulic fracturing. Given that arranging electricity generation from wind and solar energy is more complicated still, then we may indeed see a greater variety and complexity of work roles emerge in our economy as the renewable-energy industry rises and its influence spreads, meaning more job opportunities. But these jobs also tend to pay less than oil-, gas-, or coal-related work, thus the probability of lower income outcomes. Renewable-energy technologies also entail lower energy return on investment (EROI); thus my expectation that they will have a weaker impact on GDP growth rates than was witnessed during the oil age.

On the contention that renewable power will lead to less inequality in the economy, I have serious doubts. Higher-order resource-acquisition strategies in ants do seem to lead to more job opportunities or task assignments, but this shift does not lead to less inequality or less social stratification within their societies. We shouldn't be surprised to see the same outcome from a renewable-energy transition, especially as cheerleaders for wind and solar insist that we must maintain our march toward globalization and remove remaining barriers to trade and investment. Victor Yakovenko argues convincingly that this will result in further declines in living standards in the developed world and greater economic inequality based entirely on the natural tendency toward entropy in complex decentralized systems.

There's also a cost-benefit relationship at work, which the aphid-ranching ants, of course, aren't aware of but scientists are. Just think of the trade-offs inherent in human agriculture and dairy farms. For starters, farming is very hard work. Even with highly mechanized operations, farmers work long hours, and the work is hardly glamorous. Farmers and ranchers face risks associated with fickle weather and fickle markets (as I write this, many dairy farmers in the United States aren't doing so great financially), not to mention the potential for serious injury on the job. Human agriculture in general also suffers from a tendency toward monoculture, making huge swaths of crops susceptible to diseases or insects specialized to destroy that single crop. Dairy farmers aren't immune to this effect either, given that they typically feed their cattle the same thing day in, day out. But the benefits are obvious: huge volumes of reliable, storable food supplies to support a massive population (and I get to purchase cheese from France). This additional vulnerability, however, entails additional effort to protect crops and herds from pests and disease.

So it is with the ant ranchers. Their populations grow larger, and they may even live longer, but these evolved advantages don't come for free. "The cost for the ants is that they need to monopolize, collect, transport, and pass honeydew to their nest mates, which involves morphological and behavioral adaptations," according to one paper. "However, the biggest cost is likely to be that associated with being dependent on aphids for fuel for foraging."[23] In other words, these ants may assure greater chances of survival for the aphid species they are particularly fond of, but they don't actually breed the aphids as we breed our cows, pigs, sheep, and other domesticated farm animals, not deliberately anyway. Thus they've evolved a unique means of survival and of ensuring themselves access to an adequate and reliable food source, but because aphid ranchers don't directly "grow" this food, per se, they are still, to a certain extent, hostage to the natural order of things, leaving them vulnerable should some factor in the environment change in a way that's detrimental to aphid survival or reproduction. If it happened to the ancient Norse settlers of Greenland (the theory holds that climate change rendered their cattle herds untenable on that island, forcing an end to early European colonization there), then I suppose this sort of calamity could happen to aphid-ranching ants as well. Yet they forge on regardless, content to be a little ahead of the forager-only ant colonies but far behind fungus-farming ants like the leaf-cutters. That must be some really good honeydew.

* * *

The parallels between humans and ants have long been recognized and remarked on. Ants crowd themselves into bustling cities filled with millions of individuals, just as we do. They churn the earth inside out to build structures more suitable to their needs, as is our habit. Their traffic patterns snake out far beyond their central hubs so that they may access and retrieve resources to be carried back home for final consumption. They store and distribute food. They can even grow their own food, in the case of fungus farmers, or stimulate its growth, as the aphid-ranching species do, engendering even greater success for their populations, leading to greater social complexity and facilitating greater life expectancies in the bargain. All these developments are familiar to students of the rise of humankind. Many animal species display remarkable social-organization skills and attributes as well, but only the ants (and termites perhaps) seem to do so on such a massive scale, one fairly comparable to the scale of population concentration and the levels of resource extraction seen in human colonies. Hence the frequent comparisons of ants to humans and humans to ants.

Dig deeper and there are other reflections, clues as to how natural processes driving ant social evolution are probably driving human social evolution as well.

"Clustering" or "clumping" appears to be another rule of nature, though not necessarily a universal one. The patchiness of resource availability on Earth's surface encourages patchy distributions or concentrations of life. This is almost certainly one factor that drove ancient wasps living well over one hundred million years ago to begin clumping themselves into hives that eventually evolved into the massive ant colonies we see today. In lower population sizes and densities, animals will sometimes wander from patch to patch, testing each one for resource richness. As their numbers increase, they stick around the choicest patches in order for each individual to secure access to these needed resources. The populations of these clusters then increase to the point where the cluster itself becomes the attraction. Has our tendency toward urbanization really been any different? I think not. Looking at it objectively, it would seem reasonable to argue that our transition from wandering hunters and gatherers to what we are today likely followed a similar, if not identical, natural course (at least when considering the rough outline or parameters). And as clustering eventually led to the invention of farming among some ants, I believe it led to an agricultural revolution within some members of our own species as well, and we know how that turned out: the farmers and ranchers overtook the hunter-gatherers completely, and today agriculture is now by far dominant. Whether the fungus-farming ants will eventually come to totally vanquish foraging ants and conquer their own ant universe remains to be seen.

At any rate, clustering or clumping (I prefer "clustering," but either will do) explains urbanization—why humans concentrated themselves into massive settlements. Resources are undoubtedly easier to come by in the cities than in the countryside, and their distribution is far more efficient. For example, we know that human energy consumption is more efficient with urban crowding than with a more widely dispersed population, a fact that's spelled out in statistics for urban versus rural energy consumption per capita. Early human settlements began clumping around particular patches of resources for direct extraction and consumption. Over time these clumps offered value to humankind in and of themselves: defense, cooperation, sharing, and so on, but above all, better efficiency. For humans, employment is also a key necessary resource; thus people flock to the cities in search of jobs, as they are in greater abundance in the larger clusters compared to the smaller, less crowded settlements. We rush into cities despite knowing full well that the trade-off is a higher cost of living, more crowding, and more stress; we do it anyway because it's easier to get a job. And all this must be worth it, because today the cities are still attracting more and more people, while much of the world's countryside continues to empty out. This means our human clustering pattern is an ongoing process, under way even as you read this. Thus we can see the natural rhythms that explain how and why three hundred million ants crowded themselves into one small stretch of northern Japan and why tens of millions of humans are squeezing themselves into small corners of their islands and peninsulas and continents. This process appears to be nearing its end in light of the rapidly falling birthrates and consequent slowing rate of population growth driven by density dependence, but I foresee that this dynamic will play out for several more decades.

"Emergence" is another idea I wanted to share with you, a concept that Deborah Gordon arrived at via her studies into ant colonies. On only a brief look at how ants are organized, one could be forgiven for assuming that a top-down hierarchical structure is at play: a queen with her royal guard ruling over thousands or even millions of obedient workers. But the reality is a far cry from this simplistic picture. The queen doesn't rule anyone; at best she's a cow, a vessel for breeding the next generation. But the workers aren't necessarily conscious, deliberate dictators either. The system just emerged that way. And workers' roles are not fixed or firm. They change given the needs of the colony and even the age of the individual ants, and these changes in behavior are not determined by individuals or a board of directors; they emerge organically and spontaneously as a result of the millions of individual ants interacting with one another on a constant basis. The result is what we describe as the form and function of the entire ant colony as a whole. As ant expert Gordon explains, "Like many natural systems without central control, ant societies are in fact organized not by division of labor but by a distributed process, in which an ant's social role is a response to interactions with other ants. . . . [T]his social coordination occurs without any individual ant making any assessment of what needs to be done."[24]

Gordon doesn't say so, but I will: human societies almost certainly come to arrange themselves along similar lines. We may appear to be organized according to the classic division-of-labor model, top-down designed and driven, but dig deeper and we may find our reality to be rather different. The members of our colonies can change jobs as well; oftentimes they are forced to change their roles based on the changes that occur in our global economy/ecology. We end up in our careers not necessarily because we were instructed to do so (most of the time) or because we all pulled some ticket out of a hat, but as a consequence of our countless separate and unique interactions with the individuals all around us, interactions that occur in a constantly changing environment and reality. How many of us actually deliberately chose our career path, and how many of us merely stumbled into it? I'm not saying individual choice is never a factor—of course it is. Still, oftentimes we are enticed to our career choices or employment goals by clues and cues received from our neighbors and our environment. With ants these involve chemical signatures. For us we can point to the help-wanted ads in newspapers, recruiting agencies, pay scales, parental advice, career counselors, peer pressure, competition from siblings, and various other social indicators that give us hints as to what specific tasks are in popular demand at any given point in time or to the career choices most suited to us as individuals. "Emerging" from all this is what we call the world economy and the labor roles inherent within this greater human economy/ecology.

Perhaps emergence explains the general differences between democracies and dictatorships in our various national political patterns. Notice how the more democratic, less centrally organized political orders tend to rule over much more complex economies in which individuals consume a greater volume of energy per capita. Decentralized control is more attuned to the needs of complex social orders, as Gordon teaches, and it is more stable and enduring. Despotic states have simpler economies with poorer populations—the thinner diffusion of energy consumption may lead these societies to establish authoritarian rule (or the excess concentration of energy consumption into a smaller number of hands results in one). There are exceptions, of course (Saudi Arabia, Kuwait), and some states are in the midst of a transition (China), but this model I'm proposing seems to fit the pattern that we actually see. Either way, it stands to reason that our politics, along with our economics and labor force arrangements, are all emergent, just like ant social orders. It's a better argument, in my humble opinion, than assuming that multiple countries and cultures on different continents organized themselves in top-down, rigid fashions and somehow all ended up looking similar.

"Clustering" is well established in the scientific literature as it pertains to wildlife population dynamics; it's by far the most common animal distribution pattern seen in the natural world, and apparently we are no exception. "Emergence" is a rising, but extremely compelling, concept that completely flips the old and tired division-of-labor concept on its head. The third concept I've introduced in this chapter is more of my own making, though I have no doubt that women and men much smarter than I have already proposed it before. The idea is that greater societal complexity arises from higher-order resource-acquisition strategies and complexity, and not the other way around.

This idea holds that more complex social arrangements and hierarchies witnessed in the insect world and elsewhere in the animal kingdom can be explained by the level of complexity with which a species tends to extract, distribute, and consume the resources it needs or desires from nature. Greater social stratification and hierarchical organization arise from larger population sizes, for sure, but things seem to get really interesting as animals evolve even more complex yet efficient strategies of resource exploitation, and this is most evident in the case of the most radical and effective technique of resource acquisition known to this planet: agriculture. Growing your own food is undoubtedly more efficient and effective for species survival and reproduction than relying 100 percent on nature to provide your needed sustenance for you. But because agriculture is more complicated and requires more work, greater social complexity and ordering seems to result, eventually leading to a greater variety of hierarchical social layers and social stratification, as we see in nature—and, it seems, as we see among our very own human societies. Farming ants' societies are more complex than those of foraging ants, as the science seems to agree. Ant ranching also entails greater social complexity and stratification. So it must be with us as well: just as the cities likely gave rise to agriculture, and not the other way around, agriculture probably gave rise to ever greater social complexity in humans, further propelling our social and technological evolution forward. More complex energy extraction and consumption resulted in still more complex social roles and hierarchies. This process should continue as societies shift energy-generation strategies away from nonrenewable fossil fuels to renewable resources. If we ever succeed in making fusion energy reactors work, then we should see a double bang for the buck: an even greater variety of job roles and employment options coupled with higher incomes and standards of living, thanks to the far greater potential EROI of fusion.

What goes for ants goes for humans as well. That farming and ranching led to greater human social stratification is not a new idea, for sure. The difference is that scholars have long held that this process is unique to humans. In fact, it is not—this type of pattern is evident in the insect world as well; thus, it stands to reason that nature is driving these changes, not our innate human ingenuity or cleverness. In this vein we are remarkable but hardly unique. If we could teach chimpanzees or beavers or koalas to farm or herd cattle, then we shouldn't be too surprised to see them begin organizing their societies along lines familiar to any economist, anthropologist, or entomologist. Teach them to drill for oil, then look out. The rules of nature will have dictated the outcome, not human will.

However, as I cautioned earlier, we must be careful not to take these analogies too far. And at the end of the day, that's all they are: analogies. Ant societies are a lot weirder than ours, and I don't see us evolving toward some of the more extreme adaptations that now dominate the ant universe anytime soon. That's a good thing.

Ants do not exhibit the extreme division of work labor that earlier researchers and philosophers long assumed they did, but they do show an extreme division of reproductive labor that is alien. Only the queen will ever reproduce, and her children comprise the entire colony. Her daughters, the workers, are all sterile and will never bear children of their own. The males are born only to fertilize the queen, and once they accomplish this, they die, contributing absolutely nothing else to a colony. Because only the queen reproduces, the workers will dote on her children just as if they were their own, caring for the brood en masse regardless of relation or degree of fidelity. Some ant species will even care for and nurture the brood of other ant species totally unrelated to any individual in the colony (usually after raiding another ant colony and kidnapping their children— seriously). OK, that last bit has parallels in the human world, but the rest laid out above is far and away beyond our everyday experience. And the bizarre ways they've evolved their reproductive strategies aren't always advantageous. Though ants are very successful, this excessive concentration of reproductive responsibility in just the queen leaves ant colonies extremely vulnerable as well. When the queen dies, the entire colony dies.

I am fairly certain that there are no cities or countries on Earth where humans have also organized their reproductive regime in this extreme fashion—not even in North Korea. If one does exist, then I don't know about you, but I certainly wouldn't ever want to visit.

5

COLONY DISORDER

Economists and central bankers on practically every continent are growing worried again,[1] and they want us to be worried with them. They're worried that world's economy is slowing down and a great stagnation appears to be gradually settling in nearly everywhere. The central bankers' old tricks—low- to no-interest rates—just don't seem to be working this time, and new monetary tricks like "quantitative easing," whereby central banks buy up mass volumes of government and market bonds in an effort to expand the money supply, don't appear to be doing any good anymore either. So they're demanding that governments open the spigot on spending, ideally on infrastructure or other projects that create short-term jobs and demand for materials and services that can inject fresh cash into the economy and perhaps defibrillate this thing. They need the public's buy-in to make this happen. As two reporters recently put it, central banks in the United States, Europe, Japan, and beyond insist that "fiscal policy is needed where monetary policy has failed."[2]

Even before the outbreak of the novel coronavirus pandemic, the U.S. government tried to oblige worried economists to a certain extent, responding to their cries not with stimulus spending but instead with a tax cut. That worked for a while, until it stopped working again. Beyond this limited measure, the fiscal stimulus that central bankers hoped governments would be unleashing may anyway prove dead on arrival, given the gargantuan volumes of debt governments have already accumulated trying to stimulate past economic down cycles, not to mention the new debt piled on in response to the pandemic crisis. Even if they could turn on the spigot, there probably isn't very much water pressure left in those pipes. So the pundit class and the economists they lean on so much for advice are all collectively biting their nails and fidgeting in their seats, unsure of how best to move forward from here.

How about we think a bit more creatively on this topic for a change. Is general economic stagnation, inevitable or not, truly the emergency that the economists are making it out to be? Just look to nature for clues, as this entire book has been inviting you to. In reality—the realm that exists outside econometric equations—nothing grows or expands forever. The universe itself will one day die. Therefore, it seems perfectly natural that the global economy, human driven but derived from nature nevertheless, will also cease its growth pace someday, and it may even contract some-what in the future. Why should this be so bizarre or raise such alarm bells? This is how everything else works in the natural world, per the laws of physics and nature.

How should we proceed differently considering this possibility? If energy return on investment (EROI) is central to growth, then the answer is to find a new source of energy to power our global economy, one that yields truly amazing returns. A great candidate is nuclear fusion energy of the sort being developed and experimented with in southern France. Fail-ing that, the answer could be that we should stop trying to fight our way out of these laws altogether, stop looking for loopholes or ways to thwart them, and instead find new and imaginative ways to work with them or within the boundaries they impose on us. Maybe our human economy/ ecology is only now finally catching up to this reality.

I know this possibility is rather terrifying to government leaders and their economists, some of whom may be reading this right now, but fear not, talking heads—there is still time. Rest assured that for now the global economy will resume its expansion, albeit a lot more slowly than it used to. The world's population will continue to grow, just a lot more slowly as well. We don't have to see these twin phenomena as a looming crisis. Even famously birth-dearth countries like Japan, Italy, South Korea, and other advanced, mature economies are still managing modest rates of economic expansion and generally high levels of welfare and prosperity even as their populations slowly but steadily decline.

Let's now return to that other argument, regarding human population growth and the contention that we're having far too few babies to sustain economies. Strangely enough, even though the world's total human population is still very much expanding—and should continue to do so for quite some time, out to at least 2050—economists, pundits, and other experts are nevertheless instructing us all to be alarmed and afraid and stressed and worried and fretful about a looming and dangerous demographic "crisis" racing at us at full speed. The number of children we are producing is far too low, the talking heads warn darkly, and if we don't want to see public pension systems collapsing or economies falling into total disarray or other terrible, frightful consequences, then we must all pitch in and "do it for [insert name or cause or country here]!" they insist. Quite the aphrodisiac, that argument.

Opinion columnist Ross Douthat of the *New York Times* pins blame for this "crisis" on our selfish, decadent Western ways, comparing our new small-family norm to China's erstwhile harsh population-control regime, which Westerners are also somehow to blame for. Communist tyranny and zealotry and misguided moral compasses are at fault for this "underpopulation bomb" about to blow up in all our faces, he warns.[3]

Don't you feel guilty? Me neither.

For starters, it's curious that this self-appointed expert and others like him are pulling their hair out over something called a "population crisis" supposedly under way as we speak—in China of all places.

China. The Middle Kingdom. You know China, right? Where 1.4 billion humans live, most of them crammed into just the eastern third of that country. Say that out loud if you must: one billion four hundred million people. That's roughly quadruple the population of the United States, all squeezed into an area not much larger than Alaska. Douthat and many, many other like-minded commentators allowed to regularly publish their views are now arguing that this is an unacceptably low number; what China and the United States and virtually every other country really needs right now is far, far more people, not fewer, they insist.

But now that I've been afforded a rare opportunity to publish my views, I have to interject. My retort: What population crisis? Earth is plenty populated already, with well over a billion more people on the way. Have folks on the opposite end of me in this debate ever even been to China? I have. It's where I learned to hone my hatred of crowds. I don't recall seeing any wildlife there even once, though it must exist, at least in some of the less crowded parts.

What these people ultimately fear, for reasons that I still don't quite understand, is an end to eternal population growth, just as they fear an end to eternal economic growth. But here, again, there is plenty of time for us to adjust our attitudes and get used to it. Worldwide population decline is still several decades out, so there's hardly any urgency. And, unfortunately for Douthat and his friends—but very fortunately for the rest of us—this is very much an inevitability, as I think I've convincingly argued in Chapter 1.

Again, population stagnation, like economic stagnation, is perfectly normal and natural as reflected by the natural world that surrounds us. The animal kingdom holds the answers, teaching us that with overcrowding and the resulting stress, fertility decline is a guaranteed outcome. It happens with deer, but you don't see the birth-crisis-panic crowd shaking their fingers at the white-tailed deer living on Shelter Island, New York, chastising those animals for their low reproductive rates compared to deer populations elsewhere. Ocean mammals, birds of prey, and even insects exhibit this pattern, but I doubt very much that these prophets of doom spend any time worrying about the fertility rates of whales or osprey or even fruit flies in a lab. Ants are the ultimate demonstration of density dependence, and yet the panicking pundits are not yelling at anthills for their selfishness in allowing only the queens to reproduce. So why should they concern themselves all that much with human population trends? But concern themselves they do, regardless of how much we appeal for a more realistic approach, so I'm afraid their allies in the economics professions will continue to advise governments to adopt pointless policies to combat it, like the billions of dollars the South Korean government has already wasted in a futile effort to boost that nation's fertility levels. Other similar measures will also fail eventually.

As you've probably guessed by now, I disagree vociferously with these overblown worries about how many children women are currently giving birth to—which is, to reemphasize, the business of no one but the women and their partners. My advice is that folks pounding away at their laptop computers, demanding governments do something about this rising trend, relax instead and perhaps find something else to write about.[4] As I explained earlier, animal populations rise, plateau, and fall, and this pattern is entirely consistent with natural laws. We are now apparently finding ourselves forced to comply with this natural order as well (well, not the fall part, at least not yet—as I mentioned earlier we'll have to wait a few more decades until all the world's population begins to decline). That's all that's going on.

The only probable way one might counter this trend is to reverse urbanization. But, as explained earlier, that also goes entirely against the natural inclination of animals like us to cluster themselves into ever higher and denser same-species population concentrations. Urbanization and its consequent higher population concentration, otherwise known to animal researchers as "clumping" or "clustering," is the very thing driving our human birthrates lower. Natural scientists would instruct the economists that this process is entirely natural and, in many ways, inevitable for large land mammals such as ourselves.

The other so-called problems headlined in this book appear to be derived from the natural order as well. Anthills experience them just as much as our cities do.

Declining economic returns driven by falling energy return on investment may be an inevitable result in a system that relies primarily on a limited set of energy sources. Falling EROI acts the same in systems as within the biology of organisms. Inequality is also a natural consequence of systems ordering themselves rather than being ordered by an outside, external agent. These rules of nature almost certainly drive basic human macroeconomic outcomes: we grow more complex, then more crowded, then more energy efficient, until eventually our EROI hits a ceiling, the crowding that worked so well for us at the outset starts to work against us, and the inherent "metabolism" of our economy/ecology slows and slows and eventually grinds to a halt, as do population growth and settlement-size increase.

But what probably isn't at all natural or in tune with the natural order of things is what ultimately emanates from this ongoing macroeconomic phenomenon: pollution, mainly coming from our cities, which by some measures accounts for at least 80 percent of global gross domestic product. If we have all naturally ordered ourselves into colonies, as is the pattern seen with other mass-colony social animals, then the massive volumes of pollution and waste we generate daily could be pointing to some glitch in the matrix or the emergence of colony disorder, as there's nothing else quite like it anywhere in the natural world, at least not on this planet.

Clumping or clustering appears to fit with the rules of nature, as noted earlier. But human cities seem unusually unnatural in the sense that, though they are more energy and resource efficient than the alternatives, they are still enormously polluting and wasteful. Ecosystems and animals exposed to our pollution and trash suffer greatly for this, but so do we. Urban air pollution, for instance, is a persistent challenge, and not just for developing nations; the United Nations Human Settlements Programme (UN-HABITAT) finds that 49 percent of developed-world large cities do not meet basic air-quality standards. Studies into the persistent problem of air pollution have concluded that some 96 percent of humanity is exposed to levels of particulate matter in the air that exceed limits that the World Health Organization considers safe.[5] Investigations into air pollution in China have found the smog in its cities to be so bad that it literally blots out the sun, reducing the efficiency of urban solar-power projects by at least 12 percent.[6] The World Bank estimates that the planet's large cities generate more than two billion metric tons of solid waste per year, and it expects that volume to increase by nearly 70 percent by 2050 compared to 2016.[7] And don't get me started on water pollution. The form most in the news these days, the subject of regular headlines everywhere, is the huge volume of plastic materials that wash out from cities and into the oceans, about eight million tons each year, according to the International Union for Conservation of Nature. Plastic trash has been discovered at all depths of the sea, from the surface to the bottom of the Mariana Trench. Its origin is our cities.

In nature, by contrast, basically no waste is generated whatsoever. Anthills do produce landfills, but these are entirely organic and completely biodegradable. There is even solid evidence that anthill landfills actually enrich and rejuvenate soils. You definitely can't say the same about human landfills, piled high as they are with toxic and long-lasting refuse. Thousands of years from now archaeologists excavating our garbage piles will find plastic and metal containers with still-legible labels. These future archaeologists will probably have to wear protective clothing during their digs.

Anthills produce no air pollution, as well, and any water pollution they might generate is so negligible as to be easily remediated by the immediate surrounding environment. Indeed, you can say that the ants' cities seem to have evolved to be minimally impactful on their environments. Why haven't we learned this trick yet? You certainly can't argue with a straight face that New York or Lagos or Jakarta is minimally impactful on its surrounding natural environment, whether we're talking about the air, ocean, freshwater, or soil.

Ant cities and settlements can disrupt other animals' habitats and behavior patterns, no question about it, but at nothing close to the scale of disruption that our cities accomplish. Recent studies by independent researchers and UN-affiliated scientists have confirmed that the spread of human settlements is the leading cause of extinction of species,[8] exceeding all other likely causes, including global warming. It's entirely possible, maybe even likely, that the spread of anthills across the globe pushed some species to extinction as well, but that process played out over an enormous span of time, with these changes proving so slow as to be imperceptible, and such extinctions probably resulted from multiple other causes and are not the ants' sole fault. By contrast, extinctions driven by human settlement patterns are obvious and unmistakable, occur rapidly, and are thus easy to witness and record in real time. An international scientific panel fears that some one million species may become extinct in the coming decades, annihilated mainly by the spread and impact of human habitation.[9] In Houston, where I lived for several years, that city's rapid rise—it's on track to become the third largest in the United States probably around 2030—is already pushing at least one species of toad and a species of prairie chicken to the brink, and those are just the two examples that I'm currently aware of.

Some observers believe that our cities need not always be the main source of our environmental problems, arguing counterintuitively that they may one day prove the solution to them. Concentrating people into crowded cities lessens the need for physical space per person, for instance, potentially leaving more areas to wild animals and plants. There's pretty good evidence that this effect is under way right now in Japan. As the Japanese population concentrates more and more into the cities, rural Japan is emptying. This appears to be allowing wildlife to expand their ranges for the first time in a very long time, after many centuries in which wildlife habitat only contracted. There are occasionally rare but deadly consequences unfortunately: every year small numbers of elderly people remaining in the countryside will sadly fall victim to bears, usually while venturing just beyond town limits foraging for wild vegetables. Naturalists in Japan have confirmed that bears in the country are gradually expanding their range: "The mountain settlements are losing human population, and at the same time the bears' habitat range is expanding," as Yoshiaki Izumiyama, a wildlife manager in Akita, told me in 2018. "It's not that there is any rapid increase in the bear population."

Along with bears, other animal ranges are poised to expand as a result of urbanization, as will the populations of a wide variety of plant and insect species long pressured by human encroachment. If human settlement expansion in rural and urban areas pushed animals out and contributed to catastrophic levels of wildlife population decline, then it only stands to reason that greater human population concentration and a shrinking of area occupied by some settlements will eventually have the opposite impact. That doesn't seem like such a bad thing to me, though something should of course be done about rising incidents of human-animal conflict resulting from this trend.

Cities are also more energy efficient. With greater population concentration, it becomes easier to distribute both energy and other resources to a population: just think what it takes to stock gasoline stations situated hundreds of miles away in rural areas versus on every third corner in a city. Big cities in developed countries are more likely to have robust public transportation systems (at least outside the United States), efficiently moving millions of individuals around each day with much less energy expended per person compared to small-town transportation dominated by passenger vehicles. And some point out that, though they are immense sources of pollution and runoff, cities concentrate pollution sources into smaller geographical areas to the point that mitigating or managing it all may actually be easier. Cities become, in a sense, point sources for pollution, and remediation only requires that policies and technologies be put in place to plug these sources.

And there can be no doubt that some of our greatest innovations in energy and technology arrived from the laboratories housed in the cities. Thomas Edison would probably have found it much more difficult to pull off his electricity revolution from a basement in central North Dakota; instead, he achieved his feats from the urban heart of New Jersey right across the Hudson River from the hustle and bustle of Manhattan, where his funders were based. "Cities are concentrated centers of production, consumption, and waste disposal that drive land change and a host of global environmental problems," argues one team of researchers. "Our hope is that cities also concentrate the industry and creativity that have resided in urban centers throughout much of human history, making them hot spots for solutions as well as problems."[10]

I'm actually going somewhere with all this.

If natural laws are driving various human macroeconomic behaviors, then why is the general rule of "no waste" in nature somehow bypassing our global civilization? Our cities are enormously wasteful and polluting; anthills are not. Perhaps we are evolving toward this eventual outcome, only it will take far longer and require more patience. For the impatient, maybe it's possible to capture lessons from the anthills and from nature to help guide us more quickly toward to this ideal goal.

How might our cities become more sustainable and in tune with their surrounding environments—including the land, air, and water—achieving the "no waste" city so to speak? Perhaps our cities can, like anthills, be engineered to be much more "natural" in the sense of having minimally disruptive impacts on the environment, although they have already done a great deal of damage.

The ants on Earth have achieved massively populated settlements that undertake huge operations of resource extraction, distribution, and consumption, as have we; yet they generate no permanent waste and leave virtually no trace of their activities when they are gone. Perhaps we need to look at bit more closely at how anthills are organized and managed to accomplish this feat, if only to see if we can glean further lessons from ants' many millions of years of additional experience in these matters.

* * *

Though the ants probably aren't as dumb as we think, they aren't exactly rocket scientists either. One of their greatest follies in engineering their cities is one we engage in all the time: many (though not all) ant species often attempt to build hard, fixed structures and settlements within an environment that is constantly in flux. They'll build anthills only to see them washed out by floodwaters or settle and build in areas that are lush one year but then bone dry and barren the next. Sometimes they mistakenly build their settlements in the direct path of animal migrations, only to see their homes trampled and destroyed not too long after completing construction. Much the same happens with homes built in Houston or Hawaii: developers will put them up right in the middle of obvious flood zones or on top of what is clearly a young lava flow created by a nearby volcano just a few decades prior.

A better strategy would be to build temporary, more flexible settlements and accommodations that can more easily adapt to the fluctuating conditions that the ants find themselves in. But they tend toward hard, fixed cities in the first place for the same reason we do: to secure better access to resources that they need to survive and thrive and to provide for a more efficient portioning out of these resources to a very large population. Relying on mobile settlements is a great strategy for nomadic herders and some categories of hunter-gatherers, but it would be very difficult and, honestly, most aggravating for us to make our agrarian-centered civilization work if we kept constantly moving our cities and roads around annually or monthly. We grow most of our food in fixed places, so it makes perfect sense to distribute and consume it all at fixed locations as well. And even human hunter-gathering clans will build fairly permanent settlements if the surrounding grounds are particularly rich in game and foraging opportunities. That tendency toward clustering is at work, and, as I've already argued, it is probably what drove early hunter-gatherers to develop agriculture in the first place.

Thus the aphid-ranching and leaf-cutting farmer ants will expand and develop not only permanent (from their point of view, factoring in their very short life spans) anthills but even permanent "highways," which for ants are chemically scented paths that grow more established and easier for newer ants to detect the more they are used. These highways take the ants to and from their resources, though they can also be washed out or blocked by falling debris or what have you.

As with most adaptations, this tendency toward fixed dwellings isn't cost-free. There are also limitations to where anthills can be sited and how large they can be, given that they are built in a constantly changing environment that imposes certain restrictions on animals. "A fluctuating environment can impede the growth of permanent structure," as a few researchers pointed out many years ago. "Moreover, long lasting structures lack the plasticity needed in a fluctuating environment and high rates of development of new rigid structures cost energy."[11] These costs must be worth it, though we should be clear that not all ants adopt this strategy; army ants, for instance, are nomadic and don't build anthills at all; rather they form temporary "bivouacs" with their bodies by grabbing onto each other's legs.

Ants that do go the route of building fairly permanent settlements pursue strategies to cope with natural fluctuations and variable unpredictability, just as we do. We build dams or dikes to control flooding, erect storm shelters in case of tornadoes, and practice preventative maintenance in the battle against corrosion to keep our bridges from collapsing (though not always successfully, it should be said). Ant societies don't delve into this level of complexity; their adaptations typically revolve around food-stock management, or keeping enough food in reserve in the event of lean periods. "One possible reaction of the society is to try to eliminate external fluctuations . . . with homeostatic mechanisms," as an earlier-cited study notes. "The colony tries to minimize the external fluctuations by appropriate behaviors (stock management, temperature control)."[12]

So again, ants and humans both pursue their settlement-building strategies based on the same choices and trade-offs put on offer by nature. The environment is in a constant state of change, so they can either change with it via temporary, mobile living arrangements, or they can build a more fixed community but then be forced to constantly exert energy to prevent and adjust to these guaranteed-to-come environmental fluctuations. Most ant colonies choose the latter option, as we do; it's the one that makes the most sense given our resource-extraction and consumption strategies, now made requirements by dint of our enormous human population.

Of course, many ant species do move their settlements or relocate their populations to new areas to build anew, just as we humans will occasionally do. These days this occurs much less frequently than previously in the case of humans, but we can still visit the remnants of ghost towns that were once thriving mini-metropolises barely a century ago. We will also build new cities on top of older, deceased ones, as in the case of Mexico City, which was constructed on top of the ruins of the Aztec capital Tenochtitlan. Ants will do the same. Forging new paths to new territories can follow a somewhat similar process for both ants and humans. Oftentimes ants will dispatch scouts to assess the fitness of new potential colony sites, somewhat akin to how the European aristocracy did during the early days of Atlantic and Pacific exploration and colonialism, only ants lay chemical trails designed to lead their cohorts back to these new nesting grounds, whereas Marco Polo and Christopher Columbus relied on vaguely accurate maps.

But in building new settlements, ants can approach things a bit differently than we humans tend to in these more modern times of ours, at least early on.

Think of Singapore and Dubai. These are both very new cities, and although built in vastly different environments—Middle Eastern desert versus Southeast Asian rainforest—from a distance they don't actually look all that different from one another: they both comprise a central business district dominated by distinctly similar high-rise structures surrounded by a more sprawling assortment of lower-rise buildings containing both business and residences, all connected by a street and highway network trafficked by identical-looking vehicles and serviced by port facilities receiving identical-looking cruise and container ships. The picture remains much the same when you move out in opposite directions from the equator. Use your favorite internet search engine to search for images of London and Johannesburg or Sydney and Vancouver to see what I'm referring to.

Anthills are certainly a common and even dominant pattern among the world's ants, but not the only one. For example, a study of the red ant species *Myrmica laevinodis* found that its members early on adopted distinctly different settlement construction patterns depending on the environment they happened to find themselves in. In drier and open areas exposed to intensive sunlight for much of the day, the ants dug their nests deeper into cooler ground and away from the hot surface. However, these same ants built shallower settlements in the presence of abundant tree cover and shade, saving energy normally required for deeper excavation. And when there was a lot of leaf litter or debris, these red ants took advantage and spared themselves the effort of digging entirely. "The greater number of trees in a given area, the smaller the part of the nest's structure performed by the ants themselves," noted the study. "To a growing degree they make use of natural shelters, such as bedding and undergrowth. Eventually—in forests with very thick bedding or very rich undergrowth—the nests become reduced to mere hiding places under a layer of dried leaves or in the moss."[13] The ants adapted their building habits to different temperature regimes as well. What's described above only applies to nests built in warmer areas or in the summer months; the ants were found to "dig in" for the winter.

In addition, the shallower, simpler structures, such as basic dead-leaf hut nests, were found to be more common when the ants' population sizes were smaller and colonies tended to migrate more often. But it's striking to note how, at least with certain species, ants appear careful to take maximum advantage of existing shelter options that a landscape might provide, particularly in early colony-growth periods when their populations are relatively smaller. We used to do this ourselves, in fact, but abandoned this practice long ago as our numbers swelled and swelled.

There are only so many suitable caves in the world, after all. Here again we can see how our historical human settlement patterns can match processes found in the ant world. "As the family grows nest building may require greater efforts and, consequently, migratory tendencies may decrease," the above-cited ant researchers found. "Therefore, our results in this investigation may only be referred to small families and it is reasonable to assume that penetration of the terrain is different in greater societies of the species."[14]

This isn't all that different from our own experience, right? Human "penetration of the terrain" is vastly different when our numbers hit the thousands or millions compared to just a hundred or so. In the earlier period of our history, when human communities were smaller, especially among migratory or nomadic populations, structure choices demonstrated far greater variety, and building them often relied much more on materials found in the immediate environment that were easily accessible, resulting in simple structures that were nevertheless particularly well adapted to that environment (think cave dwellings, igloos, tepees, yurts, and any other variety of temporary housing unique to a particular culture and environment). But this habit of building simpler structures is only sustainable in smaller and less densely populated settlements. As populations rise, animals that evolved engineering skills, as we did, display a tendency to build more permanent settlements along more familiar, uniform lines and in accordance with repeated design patterns. Thus the ants mainly rely upon their churned-dirt-pile anthills with their elaborate underground tunnel networks populated by thousands or even millions of individuals, and we have our cities with their signature skyscrapers, suburban sprawl, and ports populated by thousands or millions of individuals, which all look pretty much the same, even if built in far-flung places with vastly different climates and weather patterns or even on opposite ends of the earth. These built-up settlements with their heavy population concentrations seem only to grow more appealing as they get larger, at least in the human experience. It's as I explained earlier: early clustering has a basic logic in that it affords more secure access to reliable sources of food and other resources for their residents, but as a cluster expands in size, eventually it becomes the attraction itself.

Yet, as enticing as these urban enclaves of animals and humankind can be, they share one unavoidable downside: they concentrate the waste as well as the workers. Waste is dangerous: it makes animals sick and encourages the spread of illness and disease, and life-threatening ailments become much more communicable at greater population densities. In essence, as waste accumulates within a colony, it threatens to unleash a very formidable decimating factor. Thus, if animals evolve into social colonies that insist on building fixed, permanent settlements and living in them in massive numbers, then these same animal colonies must quickly evolve or invent a means of maintaining proper sanitation and organize efficient and safe mechanisms for waste management. Without proper sanitation and constantly operational waste-management systems, our cities would be unlivable. It's no different for the anthills.

How humans dispose of urban waste is a complicated business, but the basics are fairly simple. Sewage is flushed out of homes with water via an elaborate network of pipes. These smaller pipes direct the waste to larger underground pipes and sewers, which then direct the accumulated wastes to large central processing facilities. These centers encourage bacteria to break down the waste to the greatest extent possible, and what can't be broken down is then separated into solids and liquids. The liquid is mostly water that's been cleaned of the waste, and it is released back into the environment; in some cases it's even reused: the city of Los Angeles, for instance, famously treats this water further and introduces it back into the city's drinking water supply, which is smart for an urban area famous for dry weather and vulnerable to upstream drought. The solids are then disposed of via burial or other methods, or they are reused as well, sometimes as agricultural fertilizer, if the waste is found to no longer contain any harmful contaminants that could hurt human health should they inadvertently enter the food supply.

For solid waste generated by households and not human bodies, otherwise known as trash, the basics are a bit more involved. Garbage is gathered into one or more corners of a home or business, then collected and removed from these buildings or even public parks and placed on the side of the street to make it accessible to collecting vehicles. These vehicles, the garbage trucks, then pick it up and carry it to central processing centers. In some jurisdictions these are incinerators, and the collected trash is burned. In other cases the solid waste is sent out to landfills where it is buried, sometimes in a tightly regulated fashion, but otherwise not so much, as is the case in much of the developing world. In some other urban centers like the San Francisco Bay Area or the Greater Tokyo Metropolitan Area, residents are instructed to separate their trash into

recyclable and nonrecyclable materials, with the recyclable materials collected and carried away to a separate facility, where they are processed
and, ideally, turned into new products (I add "ideally" because this often
isn't the case; sometimes the recyclables are simply shipped abroad,
where they may be recycled but are more likely to be buried or simply
dumped into the environment). Other higher orders of complicated wastes
are processed according to more complicated procedures (again "ideally"). Old washing machines, refrigerators, microwaves, televisions, and
other categories of household electronic waste can't simply be left on the
curb in most cases and must instead be hand-delivered to a special processing or recycling center specialized in breaking these machines down
in order to recapture the usable metals and other materials. The same
holds for potentially toxic wastes, like old cans of paint or paint thinner,
acids, containers with pesticide residues, and other potentially dangerous
materials. Medical waste is also collected, transported, and disposed of
according to an entirely separate process as determined by health authorities.

Whatever the method, with both sewage and solid-waste collection
and disposal, the general idea is the same: removal to a distance. Collect
it and then get it out of the houses, settlements, and cities as soon as
possible, ideally as far away as possible from people and population
centers. If allowed to remain and accumulate within an urban center, it
poses a great risk to the public, a risk that only rises as the waste pile
accumulates and gets older.

Never mind the toxic-waste threat; human and solid wastes are breeding grounds for a whole host of deadly pathogens and bacteria. The
number-one killer of children in the world isn't warfare or famine or
government-led genocide; it is dirty water. UNICEF reminded the world
of this fact in a report it launched back in March 2019. The statistics
presented in it are sobering and depressing: the UN children's agency
estimates that every year about 85,700 children under the age of fifteen
are killed by diseases related to water contamination, whereas wars were
responsible for taking 30,000 young lives each year, and that was from a
study that only looked at sixteen conflict-prone nations.[15] Dirty water can
have highly lethal consequences for entire nations. For example, though
cholera had long been conquered by modern sanitation, an outbreak of the
illness in Haiti following the devastating earthquake that hit that nation
back in 2010 has to date killed an estimated nine thousand people. The

United Nations, after much stonewalling, ultimately admitted that the source of the outbreak was a UN base that housed peacekeeping troops from Nepal: the peacekeepers were dumping their sewage directly into an adjacent river without treating it, and the locals relied on this river for drinking, cooking, and laundry.

Solid waste can be lethal as well. Impoverished workers who earn a living picking through massive urban dumps seeking material to sell to recyclers are expected to have shorter lifespans than their neighbors. Improper disposal and processing of electronic waste can be particularly detrimental to human health. A health study focused on an e-waste recycling center in China concluded that children living there were at risk of impaired neurological development; it even linked the e-waste to birth defects and other health consequences for pregnant women. [16] So waste of any sort should be collected and removed as quickly as possible in most cases and, ideally, processed away from populated areas.

The ants seem to have figured this out a long time ago. They also collect and dispose of waste generated in their colonies, despite the fact that it consists of organic materials that break down easily in the environment and may even be beneficial to soils. [17] But they sort and dispose of the waste anyway because it is dangerous to them and their crops. Nonfarmer ants also dispose of their refuse in landfills, so this practice isn't unique to agricultural ant societies. In fact, nonagricultural human communities will have their own waste dumps and waste-disposal methods as well, and this practice is ancient—archaeologists studying the cultures, habits, and even diets of Paleolithic peoples often find it expedient to simply sort through their trash. Mass waste disposal is a unique animal behavior but a very necessary one for species of social animals that concentrate themselves into smaller areas in very large numbers, as humans do. The ants have simply been running large-scale waste-management operations far longer than we have.

Some ants remove wastes from their colony and collect them in a dump or landfill on the surface and away from the anthill. Other species build special underground chambers separated from the living and working quarters and dispose of their wastes safely there. Whatever the preferred method or strategy, the idea is the same as in our cities: get rid of the waste as quickly as possible, removing it from the concentrated population where it poses a great threat to the society's health.

"Hygienic behavior is an important aspect of social organization because living in aggregations facilitates the spread of disease," note A. N. M. Bot and colleagues, who researched the waste-management systems of leaf-cutter ants, the highly successful fungus farmers. The research they undertook on both above- and underground disposal methods determined conclusively that "waste is dangerous for the ants, which die at a higher rate in the presence of waste," while also finding that "waste is dangerous for the mutualistic fungus because waste in field colonies is infected with the specialized fungal parasite Escovopsis."[18] The ants studied for this particular paper were found to be exerting considerable effort to keep their fungus gardens free of this parasite. The stakes for the ants were quite high. The researchers noted some instances of uncontrolled Escovopsis parasite outbreaks, with the parasite occasionally completely overrunning and destroying colonies' fungus gardens. In those cases the entire colony would die as a result, having lost its only food supply. Consequently, "the ants allocate considerable effort to active management of waste in order to reduce these dangers."[19]

And leaf-cutter ants can generate a surprisingly huge volume of waste given their large populations, with colonies capable of numbering in the millions of individuals. A study on leaf-cutter ant garbage conducted in 1947 excavated the trash piles from some 296 underground waste chambers built in one particularly large leaf-cutter ant colony. The accumulated trash weighed in at 475 kilograms, or more than 1,000 pounds.[20]

And in the ant universe the challenge and complexity of these waste-disposal efforts scales up as ant colony populations grow larger, just as it does with humans. "The larger the society, the greater the challenge faced in waste disposal," said two scientists from the University of Sheffield in the United Kingdom, referring to both humans and ants.[21] The ants can slip up on occasion and mismanage things, as noted above when leaf-cutter ant colonies die out for failure to properly isolate their precious fungus gardens from their garbage. We fail to perfectly manage our own waste streams all the time. To this day New York City still can't handle a heavy rainstorm—every downpour overwhelms the city's sewers, resulting in millions of gallons of untreated raw sewage spilling out into the city's surrounding waters. The problem is well understood and fairly easy to remediate by simply building another wastewater treatment plant, but New York's leadership has deemed that solution cost prohibitive, and so the issue goes unresolved despite the obvious risk it poses to the health of the city's residents. I'd be curious to know if the leaf-cutter ant colonies observed to suffer the destruction of their fungus gardens by a parasite outbreak were among the heavier-populated colonies, with the ants falling victim merely to the greater size and complexity of their waste-management challenges.

The ants are also rather careful about where they locate surface landfills. The authors cited above note a positive correlation between colony population and distance of the landfill from the anthill: the greater the colony size, the farther away any landfill was situated. The ants studied were also mindful of runoff. "Waste heaps are always located downhill from nest entrances," they noted.

Also, it appears that greater waste-management complexity yields greater social complexity in ants as well. You'll recall my earlier explanation that resource-extraction complexity appears to see social animals and insects evolve a more complicated division of labor. This is the other side of that coin. For the leaf-cutter species they investigated, the waste-management division of labor "is undertaken by transporters that carry waste to the heap margins and heap workers that manage the heap."[22] The heap workers essentially churn the waste over, in an apparent effort to help it decompose more quickly, much as a gardener might occasionally mix and churn a compost pile. This ant waste-disposal team gets very little sleep as well—the process of gathering, dumping, and churning the waste occurred all day and night, in contrast to foraging activity, which picked up in the daylight and slowed considerably after dusk.

These ants also enforced a strict division between the waste workers and the foragers, or the ants tasked with finding and carrying back food. Waste workers were never allowed to forage, it appears, with the study concluding that this division must have evolved as a means to prevent the waste transporters and sorters from infecting the colony's food supply. Delivery of waste from its origin to its ultimate destination somewhat resembles the way we do things. For humans, households or businesses leave their trash outside for waste-management workers to pick up. Leaf-cutter ants work things pretty much the same way; those tending to the fungus garden gather the garbage that results from that activity and remove it to outside the fungus garden chamber, and a waste transporter later picks this load up. "Workers from the fungus garden place refuse particles on a cache just outside the dump, from which it is further transported by dump workers."[23]

It's somewhat heartening to know that both ants and humans recognize the critical importance of sanitation, waste management, and disposal and that both animals exert considerable amounts of energy toward continually moving waste away from their cities and population concentrations as quickly as they can manage. Other species are much more cavalier in their attitudes to their wastes. I'm sure you've seen flocks of birds happily skipping along cliff rocks covered in their own feces. But birds are generally mobile, whereas we are not. As with human cities, moving an ant colony comprising millions of individuals gets incredibly difficult, so their denizens tend to stay put and work hard, 24/7, managing ant cities' waste for the sake of public health. But our approach to this problem is a bit more altruistic and egalitarian than the way the ants approach things, or at least in the case of some ant species.

In the Bot et al. study, the research team discovered that the colony under investigation relegated waste-delivery and -disposal functions to the oldest members of the colony. It's a cruel fate but the ultimate culmination of a long process whereby workers gradually move farther from the center of colony activity and out toward its margins as they age. "Near the end of their lives they engage in risky tasks such as foraging or possibly waste management," researchers noted. "As a consequence of this type of task partitioning, only the older and less valuable workers are exposed to infectious agents from the outside environment while the reproductive center of the colony is relatively protected."[24]

In other words, these ants considered their elderly to be expendable; thus the oldest workers were tasked with the work that most exposed them to potential illness. And as you read earlier, ants in close proximity to their waste die sooner than those that are kept away from it. Thus the colony somehow evolved to shunt older workers farther and farther out of the center of colony life, until they were ultimately relegated to waste management as their final function on Earth, given that their remaining lifespans are already very short. This seems particularly cruel to me, but then again the ants mean no malice by engaging in these practices. It's simply the way they've evolved, and they have no control over it.

Our own human world hasn't yet reduced itself to this level of cold Darwinian calculus, but there can be no doubt that the waste workers of our cities are oftentimes the most marginalized members of society, toiling away in the background almost invisible to the rest of us politicians, accountants, factory workers, lawyers, and reporters. We ignore them and sometimes even look down on them, but the fact remains that none of us would be alive were it not for their thankless work. Cities, like anthills, simply can't exist without efficient, large-scale waste management and disposal. Resource acquisition, distribution, and consumption may be the most important activity for social animal survival directly, but waste collection and disposal is right up there in degree of importance to anthill and human-city success. The rules of nature demand it, and so it is.

* * *

Until now I have failed to draw any significant parallels between the ant world and our own in a way that might lead us toward that hoped-for destination of a waste-free, minimally ecologically impactful human city. Ants cluster themselves into crowded anthills, then develop and enforce rudimentary, somewhat flexible divisions of labor. Tasks for the workers change as they grow older, but ants do not retire as many of our elderly do. Quite a lot of them may be rendered irrelevant and forced into unemployment, but the ants seem most generous in their welfare states as these ants are still permitted food and shelter free of charge. Crowded anthills also exert enormous quantities of energy to ensure clean and sanitary conditions as best as they can manage, given their immense population sizes. Waste is collected on a regular and never-ending basis, carried away from the population and then disposed of, as is the case with humans. All well and good; however, none of this shows us why anthills are fundamentally more sustainable and easier on the environment than urban concentrations.

Perhaps lessons lie with another amazing, if rare, ant behavior: solid-waste recycling and reuse.

Direct recycling or reuse of waste is exceedingly rare among the animals, social or otherwise; in fact, you almost never see it. Even for humans, recycling of even a fairly small percentage of the total of the enormous waste stream flowing out of our civilization is a fairly new phenomenon. This is because recycling entails costs and requires energy, and more often than not it is much cheaper and less energy intensive for animals to simply acquire and use new resources and materials, then dispose of the resulting waste, than to sort out ways to reuse existing refuse or waste from dumps. So it is for humans: even today, despite enormous political pressure for our cities to lighten their environmental footprints, most still do not recycle, as arranging a system for collecting, sorting, and recycling used materials can prove enormously complex and expensive. There are exceptions, of course, and in some corners of the world, it is becoming rather economical to recycle trash, especially as it represents free raw materials. Corrugated cardboard, for instance, is the most commonly recycled material in the United States because it is relatively easy to acquire and transport (given that it is lightweight), and it's inexpensive to break apart cardboard boxes and manufacture new cardboard out of this mix. The recycled boxes can even be sold at a premium over boxes made without recycling, further incentivizing this practice.

But the broader world of animals and insects seems rather averse to recycling.[25] The environment and their ecological survival strategies don't require them to recycle their trash, so they don't. "Animals adjust their behaviors in response to changing environmental conditions because the costs and benefits of such behaviors change as conditions change," as two Argentinian ant researchers put it. "The reuse of materials from waste (i.e., recycling) rarely occurs in social insects because it may imply significant health risks and behavioral difficulties."[26] Our species can easily mitigate against health risks associated with recollected spent waste because we're smart enough to figure out how, but this obviously isn't the case for other organisms. Plus, recycling involves fundamental adjustments to animal behavior, as these scientists noted above. Practices that underpinned the original creation of the social animal or insect colonies in the first place must absorb radical changes in order to incorporate the usually dangerous wastes so that they somehow play a role in that society's function, and performing these adjustments in behavior requires much additional labor and expenditure of energy. These costs in almost all circumstances generally outweigh any benefits that might accrue from adapting or evolving recycling behavior. And so, most of the time animals don't practice these behaviors: they simply dispose of their waste and move on, as we have for thousands of years.

The relatively new study out of Argentina referred to above is careful to note an important caveat. "However," the team hedged, "behaviors occur in the context of ever-changing environments, and so costs and benefits also change." In other words, in certain circumstances cost-benefit rationales can flip as a result of changed conditions in an environment, so it's entirely possible that nature could create conditions whereby recycling of discarded waste is of ecological benefit to an animal and worthy of the costs and energy exertion associated with that activity. And indeed, this does seem to occur. For sure most ants do not recycle, but some do, as these researchers noted in their groundbreaking 2012 paper.

A particular leaf-cutter ant they studied intently was found, much to their surprise, to be excavating previously discarded waste from its landfill and then reusing this waste material to repair damages to its anthill, damage usually caused by other animals. But the ants did so only when conditions compelled them to: the researchers say that this leaf-cutter ant species would recycle its landfill waste and reuse it as building material only during particularly hot times of year. Ant foraging behavior slowed when it was deemed too hot for workers to be overexerting themselves, and this left the colony short of foraging material. In addition, the higher surface temperatures threatened the health and welfare of the ants' precious underground fungus garden. Damage to the anthill meant potentially exposing the fungus garden to this heat risk, and so, facing shortages of fresh supplies due to subdued foraging activity, these ants resorted to mining their landfill for the materials needed to undertake repairs. "Thus colonies with lower foraging rates apparently use their refuse to repair mounds because this substrate requires less searching and carrying time," the researchers concluded. "Recycling of discarded materials might be an adaptive and flexible response to limit waste accumulation generated by dense societies and to confront restrictions in the availability of resources."[27]

Reading only that last line, one might mistakenly think it was written with humans in mind, but no, the paper doesn't mention humans at all; nor does it attempt to draw any similarities or distinctions between certain leaf-cutting ant recycling behavior and our own, but that quote fits our experience pretty well, I would argue. We've very recently begun making attempts at large-scale waste recycling, given our concerns over the steadily increasing accumulation of wastes, the filling of landfills, and the very visible negative harm to the environment brought about by uncontrolled solid-waste disposal; the immense volume of plastic trash floating in our oceans is perhaps the most visible consequence of our profligacy. Thus we are beginning to ramp up solid-waste recycling effort in an attempt to "limit waste accumulation," as these authors theorized for the ants. However, the other side of that coin, an apparent need to "confront restrictions in the availability of resources," hasn't come to pass for our species yet. In most jurisdictions recycling still struggles to compete with the normal, everyday business practice of making products out of raw materials, so where supply chains are still met with a relative abundance of new raw materials, one will find recycling rates to be lower, unless recycling practices are mandated outright by government decree (and such government mandates usually incorporate subsidies to recycling industry segments to ensure they can better compete economically in the marketplace with standard business practices).

However, recycling for humans generally occurs long before the material is sent to dumps—we capture it from our waste streams and move it into a different process separate from normal solid-waste disposal operations. What's remarkable about the leaf-cutter ant species cited above is that its workers took to excavating an existing landfill, finding in that dump the materials they felt were best suited for conducting nest repairs, and then reusing this waste matter as they saw fit, but only when it involved expending more time and energy seeking out additional raw materials from nature instead. Our species does not practice landfill excavation or dump mining on any kind of regular, organized basis or at particularly large scales. But we've certainly experimented with landfill mining in the recent past and continue to play around with it in some corners of the world.

A scholarly paper by Arindam Dhar of the University of Texas, Arlington, says humans first began experimenting with modern urban-waste landfill mining in Israel in the early 1950s as farmers sought out suitable refuse matter to fertilize their orchards, but these experiments apparently didn't last. Landfill mining first appeared the United States in the late 1980s to early 1990s as dump owners tried to find ways to reduce costs or their exposure to increasingly stringent environmental regulations. Landfill mining next emerged in Germany, where, as the author notes, some laws began mandating the recovery of reusable materials during companies' attempts to either clean up or relocate these buried waste piles (newer laws or regulations may require companies to move waste originally placed in unlined landfills to newer lined landfills, for example). By the 2000s experimental landfill mining had spread to the United Kingdom, the Netherlands, China, India, South Korea, and Thailand.[28]

This particular study expands the use of landfill mining to include recovery of waste for burning, so the picture painted in it is a little murky. Landfills can generate enormous volumes of methane gas as the buried waste decomposes, and this methane routinely escapes into the atmosphere. Some clever people have long ago figured out that you can capture this waste via pipes and other infrastructure and burn it to generate electricity or sell it to the local gas utility to be mixed into its natural gas distribution network. Other projects have toyed with gathering burnable solid waste as a fuel source, as was apparently intended for a project in Florida. But a desire for usable material didn't always underwrite these projects; for instance, landfill mining in certain cases in Italy apparently came about because urban areas were running out of space on which to build. [29] And the vast majority of these early tentative steps toward landfill mining involved temporary pilot projects and were found to have been inspired mainly by environmental regulations and concerns. "Nearly all of them were motivated by local pollution problems or hazard prevention," says Dhar. "Resource recovery was seldom the driver of [landfill mining] in the past, but has recently gained more importance." [30] Projects in Germany mark a particularly important turning point, he argues, as the government in Bavaria began enticing old landfill exploration and mining via subsidies.

But even today landfill mining occurs in a very limited number of cases and makes an insignificant contribution to society's overall resource needs. As with ants, the costs for us in undertaking this expensive and dangerous practice generally far outweigh any benefits that might be gained from mining landfills in search of recyclables. Human landfills are foul, and ripping them open to collect materials pollutes the air and raises a risk of water pollution as this material is now directly exposed to rain runoff. Our landfills routinely contain toxic materials. We can manage to keep workers safe from these toxic compounds but must then get rid of them again in a safe manner consistent with environmental rules, which only adds further costs. The material must then be sorted as recyclables are separated from the much larger volume of unusable or undesirable refuse, in the same way that terrestrial mineral mines generate huge volumes of waste slag in order to get at the much smaller concentrations of valuable ore.

Then again, as is the case in the ant world, cost-benefit analyses can change because our environment is constantly changing. It's entirely feasible that future resource scarcities or restrictions could change the math to see landfill mining emerge as a cost-effective and efficient alternative to raw material acquisition, and thus these changed circumstances might compel us to dive into large-scale commercial landfill excavation and refuse recapture and reuse. If this has already happened for at least one species of leaf-cutter ant, then there may be hope for us yet.

But again, for landfill mining to eventually rise to become a mainstream resource-acquisition strategy for our economy/ecology, the benefits netted from engaging in this complex, expensive, and potentially dangerous practice must well outweigh the costs of doing so. That's how it works for the ants. The same restrictions that dictate where and how waste-pile recycling can emerge in nature apply to our human economy/ecology as well. Even strong proponents of landfill mining acknowledge this fact.

"In order for landfill mining to be feasible for individual companies, economic benefits must of course outweigh the costs," as one literature review out of Sweden concluded. "So far this type of project has mainly been initiated, funded and operated by local authorities, i.e. owners of landfills, aiming to solve a specific issue of relevance for their region, such as a lack of landfill space. . . . Initiatives emphasizing extraction of valuable resources from deposits on commercial grounds are, however, something significantly different."[31]

So what additional lessons have the anthills provided to our economy/ecology so far in this chapter? Very little, I'm afraid.

We know ants tend to build fixed structures in a constantly changing environment, and this carries risks. The same holds for us. The ants do this because they are compelled to do so as they cluster near ideal resource patches, as their numbers swell, and as it becomes much more difficult for some species to coordinate the relocation of colonies when millions of individuals live in them. And higher population figures beget more uniform building patterns—hence the earthen mounds versus our skyscraper-studded cities. Building a fixed settlement means exerting constant energy to adapt to constant environmental change, thus requiring even more energy and resources and greater social complexity. These concentrations also generate huge volumes of waste especially as their population numbers increase. Waste is dangerous and threatens survival; thus ants evolve management and disposal practices, which seem to lead to further evolutions in social complexity and worker stratification for social insects.

Most promising for us in this discussion: solid-waste recycling and reuse does seem to occur in nature as well but only emerges in very limited circumstances and only when the benefits outweigh the costs.

That may help explain recycling's very limited penetration of our own economy/ecology. Landfill mining and waste-material reuse can occur as well, only it's rare and also dependent on a strict cost/benefit assessment. Lucky for us the environment can change in ways that render recycling far more beneficial than it is expensive. So there's hope: our economy/ ecology may one day reach a point where routine recycling and even landfill mining may supplement or even replace nonrenewable resource extraction as the primary driver of our civilization. But given the enormous variety and severity of our most pressing environmental challenges—ocean trash, water pollution, soil degradation and loss, and global warming—we may want to consider finding ways to help speed this transition along.

Here, I've reached the limit of my powers of analogy.

Anthill economies generate solid waste, as do ours, and ants rapidly dispose of wastes, as we are prone to. Their solid-waste dumps are sometimes scavenged for recyclable materials, in rare circumstances. But the ants actually have a much easier time of landfill mining than we do because, though these dumps potentially hold dangerous pathogens, they are entirely organic. They contain no toxic chemicals, metals, or plastic residues. Ants don't produce radioactive waste either. Anthills also generate no air pollution, unlike human cities, and any water pollution they may be responsible for is negligible. They are sometimes organic farmers or ranchers, but they have no industries, as their material needs are simple and easily met through exploitation of their immediate surroundings. That's why anthills fundamentally do not present huge environmental challenges like our cities do, so unfortunately ants have no valuable experiences in waste mitigation and management to share with the rest of us.

There is no perfect energy-technology solution for powering the world's transportation, telecommunications infrastructure, and economy. All energy sources come with pros and cons, benefits and pitfalls. Some of the most ardent supporters of fossil fuels or renewable-energy technologies will occasionally admit this fact, though much recent discussion about the world's future energy mix swings between an "all of the above" strategy and environmental activists' argument that success with 100 percent renewable energy is possible. Still, there's no question that our civilization's overwhelming reliance on fossil fuel burning for energy generation has had severe repercussions, the biggest of all being humanity's alteration of the composition of the atmosphere. Today the relatively thin envelope that separates us from cold dark space contains far more carbon dioxide than at any point since the time of the dinosaurs.

Ants can't help us solve our energy and economy/ecology dilemma, including this greatest challenge, because they don't capture and consume the same energy resources as we do and thus need not trouble themselves with the same pollution and consequent environmental challenges that we've inflicted upon ourselves. We understand the problems, but finding the solutions is proving far, far more difficult than many of us would like, and thus far nature isn't presenting us with any guideposts. But this shouldn't come as a terrible surprise; it is extremely difficult for animals to completely change the economy/ecology that underpins their very survival strategy. After all, the ants have been stuck in their stubborn ways for millions of years, which is fine for them so long as they continue to experience no real consequences for doing so.

Some of you reading this now may think the answers are obvious, but let's pause a bit to really consider our options.

Renewable energy holds many obvious advantages over fossil energy sources, in particular dramatically lower carbon dioxide emissions from renewable-energy distribution and zero emissions during direct power generation. But for all their faults, the shale oil and shale gas revolutions solved a lot of vexing economic challenges for the United States that solar and wind power haven't yet put a significant dent in. The large number of high-paying jobs fracking created revitalized communities on a scale that renewable-energy promotion has yet to match. This has not gone unnoticed in much of the nation's interior, and there's no indication yet that renewable energy can accomplish something similar to what fracking did in terms of economic rejuvenation in places like western Pennsylvania and parts of West Virginia.[32] Wages are still generally higher in oil, natural gas, and coal extraction than they are for assembling and maintaining wind- and solar-power projects. The U.S. public's focus on trade shouldn't be dismissed in this discussion either. Though renewable energy is low in greenhouse gas emissions, its appeal may remain limited among the wider public so long as the needed equipment continues to be manufactured mainly overseas, exacerbating the United States' already colossal trade deficit (and the larger a trade deficit, the slower economic growth becomes, a fact that most business and economics reporters in America are careful to never mention).

Intermittency remains an issue for renewable electricity as well. Wind is more consistent but subject to lulls. Solar panels don't work at night. The intermittency issue hasn't proven all that problematic in some of the world's leading renewable-energy jurisdictions up to now, but that could change as wind and solar projects grab more and more of their share of the grid. Proponents would counter that intermittency can be overcome by grid storage solutions like advanced battery applications, but the argument relies upon the successful realization of nascent technologies that have yet to be proven to work on very large scales.

Oil, natural gas, and coal are hardly perfect solutions either. Fossil fuels are abundant, broadly dispersed, and very energy dense. They can be obtained in large volumes with existing technology. But the cons are obvious: Extracting coal leads to great environmental damage. The casings and cementing on oil and gas wells often fail, which can result in contaminants leaching into groundwater. The risks inherent in offshore oil spills are plain for all to see. But the main downside to fossil fuels is the emissions realized from their combustion, chemical compounds that are either toxic or can trap heat in the atmosphere for an extended period, leading to the man-made global warming phenomenon.

There is also an economic downside to the fossil fuel business: the boom-and-bust cycle. Lots of high-paying jobs can be created during an upturn, sparking economic growth in even isolated rural communities that can yield real, measurable gains in a fairly short time. The pace of economic benefits realized by the Eagle Ford Shale came faster and at a volume greater than researchers at the University of Texas, San Antonio, initially expected, according to their studies. But the longevity of these good times is entirely dependent on the whims of the global commodities markets. When the busts come, they hit hard, and the job losses, manufacturing downturns, and shortfalls in government revenues are a repeating human economic/ecological phenomenon.

Renewable energy is clean and emits zero greenhouse gases while generating electricity (though emissions result from the manufacture, transport, and installation of the equipment). Renewable energy does generate employment and income, but not on the scale often envisioned by its most ardent U.S. supporters, at least not yet, and in the United States these projects depend heavily on foreign technology, dragging net economic growth lower and thus threatening broader public support outside policy circles and pro-renewable-energy organizations. Fossil fuels are abundant and can be extracted domestically. The entire chain of shale oil and gas extraction, delivery, and processing generates a large number of domestic jobs that generally pay good salaries to women and men who lack higher education degrees. However, the 2010 Deepwater Horizon oil spill in the Gulf of Mexico only too clearly demonstrated the potential environmental harms that can be realized from such accidents, not to mention gas leakages and onshore well casing and cementing failures. The risks of fossil fuel extraction can far outweigh the benefits, and global warming seems to be pushing this math ever more in that direction.

Because the earth is warming, the thermal expansion of the oceans could cause global mean sea levels to rise by at least three to six feet (one to two meters) over the next hundred years or so (dictated, again, by the laws of physics). Further, computer models suggest that the higher CO_2 concentration will likely lead to a greater frequency of irregular weather events and that regional climate patterns could become exacerbated, diverging from current experience (wet regions get wetter, dry regions get drier). These models, of course, are imperfect. But we do know for a fact that the ice caps are melting on Greenland and Antarctica, and this could raise sea levels far higher. Severe weather and regional climate disruptions could prove worse than suspected—worse than what we've been experiencing lately. What will actually occur in the future is unclear, and the farther out in time you go, the greater the uncertainty becomes, but ignorance is no excuse for inaction.

Like ants, humankind demonstrates a remarkable capacity to survive in nearly every environment where there is breathable air, from cold high latitudes to barren deserts. But surviving whatever alterations to our planet climate change may bring could prove very expensive, inconvenient, and even deadly for large numbers of humans. Much damage to coastal infrastructure is likely. Whole populations may be upended over time. Tackling global warming also serves as a good excuse to tackle early on a very real threat to global civilization: fossil fuel depletion. Fossil fuels may be abundant, but they are not infinitely so, and the world economy's overreliance on nonrenewable resources is a civilizational threat.

So we are collectively aware of the ecological damages wrought by our fossil-energy dependence and the enormous risk of continuing with the status quo; yet the solutions elude us, and nature seems bereft of any clear answers or analogies that may be of use to us in this discussion. But the ants needn't trouble themselves with our problems, because they don't seem interested in evolving to the level of energy and resource extraction that we have. The answers may lie simply in which energy technologies our human species relies on the most.

For a majority of its existence, humankind and its antecedents survived in climatically acceptable tropical and subtropical regions of the world similar to where the species evolved. By harnessing fire—energy through the combustion of plant material—humanity enabled its population to grow, and its range expanded to include much colder climates as well as temperate zones. Tapping into coal-fired power permitted the human population to boom again and cross oceans. This reality set advancing, high-populace societies clashing with hunter-gatherer tribes, with lingering consequences. Oil extraction led to another explosion in human population size, fueling urbanization and bringing us to the current situation in which we find ourselves. But the fossil fuel revolution exposed us to the risks inherent in overreliance on polluting nonrenewable sources of energy, climate change being the threat of greatest magnitude, though by no means the only one.

Now our species may or may not be embarking on a fundamental break with fossil fuels. Those chasing this goal see society and its economy transformed with widespread renewable-energy generation or perhaps a hypothetical replacement of both fossil fuels and traditional renewable energy with mass fusion and/or fission power generation. Both paths have the added advantage of potentially putting a major dent in the climate-change problem. Meanwhile, this is a fossil-fueled global economy, and coal, oil, and natural gas are not going away anytime soon. The Paris climate accord doesn't change that equation, but excitement and plenty of surprises lie behind the ever-quickening rise of renewable-power generation. And the nuclear fusion and fission visionaries may come through yet.

Regardless of which vision ultimately becomes reality, how these comprehensive revolutions will in turn impact the species, the economy, or life on Earth is anyone's guess. Meanwhile, new technologies and alternatives are worth pursuing, because revitalizing the world's stagnating economy, if it is stalling, will certainly require new energy, and ideally energy with a solid return on energy invested.

Sorry, but it appears that we are the only social animals on this planet prone to extreme colony disorder, evolving the means to wreck the very place that made our existence possible with our energy use and our resource-extraction and -consumption behaviors. The ants can offer no lessons for how to get out of this mess. I'm afraid we'll have to sort this one out for ourselves.

CONCLUSION

As I write this, a terrible disease has spread across the globe, wreaking immense damage everywhere it has landed. The response by experts, scientists, and the governments they advise has been ad hoc, haphazard, and far too late.

Here I am actually not speaking of COVID-19, the dreadful disease caused by a new coronavirus that emerged in Wuhan, China, in late 2019. Make no mistake, I empathize with the victims and families impacted by the worst human viral pandemic in modern memory, and we will all mourn for them and for the world's loss for some time to come. My wife and I highly suspect that we fell ill to the new virus ourselves. As it was spreading beyond China, we both succumbed to a strange illness, a "flu" that our doctor determined wasn't a flu, with stubborn symptoms identical to those described on television and lasting for two weeks, including a fever (which was particularly alarming for me, as I hadn't had one since age eleven) and an unusually strong, sometimes painful cough. Because our symptoms were mild enough, the authorities in Japan wouldn't test us for the novel coronavirus, so we may never know for sure what sickened us.

There will be plenty written about COVID-19 for years to come. This book aims to better explain the forces working upon our society and economy by showing how these same forces are at work in the animal kingdom. Disease outbreaks are yet another example of this reality we are subject to, so the pandemic I'm referring to above is killing animals and not us, spread not by a virus but by the chytrid fungus, which causes the disease chytridiomycosis. This infection has so far proven harmless to humans but has taken a terrible toll on frogs and salamanders. Contracted via the waters that amphibians wallow and feed in, chytridiomycosis has spread throughout Eurasia, the Americas, Africa, Australia, and New Zealand. Scientists investigating this particular animal plague have concluded that the disease has likely wiped out approximately 90 percent of the world's amphibian species as it's raged since the 1970s, pushing five hundred other amphibian species to the brink of extinction. It's considered one of the worst animal plagues ever discovered, and the mobilization against it has proven incredibly slow—though the disease was first identified in the 1980s, ecologists didn't get around to devising plans to fight it until 2006.

Like humans, animals and insects fall victim to widespread disease outbreaks all the time, and, like us, they have developed their own defenses and coping mechanisms. Most of the time these defense systems work just fine, and the animal population is left relatively unscathed. But when an entirely new pathogen emerges for which a population has no prior natural defenses or immunity, then the outcome can be absolutely devastating, as seen with chytridiomycosis and with "white-nose syndrome," a disease first identified in a cave in upstate New York that has to date killed millions of bats throughout North America.

Like humans, ants have proven remarkably successful at keeping pandemics at bay throughout their evolutionary history. I've already described the ants' careful attention to waste management and disposal, with colonies expending enormous energy to keep wastes from threatening members' health. They also keep meticulous hygiene habits and can even disinfect their anthills by spraying a natural chemical around to kill pathogens. Researchers at Austria's Institute of Science and Technology led by Sylvia Cremer have discovered another fascinating adaptation that some ant species use to guard against a potentially deadly fungal infection outbreak. The Cremer group discovered ant colonies that essentially inoculate themselves against a spreading infection. In this case healthy ants will lick and consume spores that break out on the bodies of infected members of their colony. The bacteria in their mouths and guts weaken the pathogen to the point that their immune systems adapt to it, and the ant colony survives even as more and more ants become exposed. This is, in a sense, nature's answer to humanity's enormously successful vaccination campaigns. In fact, according to one account, the early Chinese made attempts at smallpox inoculation via a method very similar to what these ants do, trying to build up immunity by inhaling ground smallpox scabs in experiments dating back to the 1500s.

Unfortunately, not all species are as well adapted and highly specialized as the ants in their dealings with disease or as clever as we humans have proven at artificial inoculation and immunization building. For instance, bats and amphibians are clearly succumbing in dreadfully high numbers to their pandemics. However, occasionally livestock and wildlife will enjoy the protection of expert human intervention, with managers and health experts attempting to limit the spread and impact of an emerging pathogenic threat to animals through a variety of strategies. The steps taken are not just similar to those used recently in defense of human populations against COVID-19; the playbook is identical, as a team of ecologists and evolutionary biologists from California, New Mexico, and London detailed a few years back. "Reducing the probability of introduction through quarantine and trade restriction is key because prevention is more cost effective than subsequent responses," they noted, referring to ways to prevent or contain a disease outbreak in wildlife populations. However, in locations already experiencing an epidemic, they emphasized, wildlife population managers should focus on reducing transmission and disease "through further quarantines, distancing, contact tracing, and other methods until treatments or cures can be found or until the population as a whole develops a general resistance to the disease."[1]

Again, we find that what holds true for animals and insects holds true for humans as well.

On a more hopeful note, there are indications that the bats threatened by white-nose syndrome may fare better than the amphibians being wiped out by chytridiomycosis. Named for the white growth that appears on afflicted animals' snouts, white-nose syndrome propagates because bats congregate in very large numbers. The white infection irritates them and prevents them from getting enough sleep, to the point where they expend too much energy that can't be replaced through nighttime hunting, causing their energy return on investment (EROI) to fall to the point where they die of starvation or exhaustion. Ecologists first noticed in 2007 that something was killing huge numbers of bats in upstate New York, and within a year they had identified the culprit as white-nose syndrome. By 2011 managers in the United States and Canada had developed action plans, a record response time considering the decades that passed before wildlife conservationists began taking the amphibian plague seriously.

* * *

Skeptical readers, including many professional economists and scientists alike, will probably find much to dislike about this book, and I can already imagine why that might be. It's not for everyone, I'm sure.

Point of fact, I am no scientist; nor am I an economist. I'm just a trained observer and science buff, one with enough exposure to economists to know fully well where their strengths and weaknesses lie and how their weaknesses and often willful blindness to reality badly hobbles their profession. The economists' immense political power and very obvious, colossal failures, suffered by all of us in recent years, have left them almost wholly discredited in the eyes of the broader public. As I mentioned earlier, the field of economics is now among the least trusted professions.[2] So, they're long overdue for a reckoning and comeuppance. This book is but one of hundreds (perhaps thousands) of volumes that attempt to better understand why they've failed us so miserably on so many occasions and why we don't necessarily need to listen to everything they say or take them so seriously anymore. But I'll be the first to admit, this book doesn't have all the answers; nor does it pretend to.

Because the ideas laid out for you in these pages are well beyond the mainstream of economic thought, the mainstream economists will almost certainly dismiss all of them. Most likely many of them will loudly scoff at this book's arguments and reject the lessons from the naturalists outright. That would be par for the course. I'm entirely anticipating and perfectly OK with this reaction. But we should take their typically dismissive attitudes as an invitation to freely dismiss their assertions and certitudes as well, especially in light of the multiple disasters they've led us into. Besides, if these "experts" ultimately can't accept the idea that some basic fundamental laws of nature and the universe might have some bearing at all on their chosen profession, then what truly qualifies them to lecture the rest of us in the first place?

Progressive-minded individuals will probably find much to dislike about this book as well. Its general outline of the way the world probably works is pretty disappointing after all, especially during election cycles in which left-wing populists and self-described democratic socialists are often leading contenders in the polls. Economic stagnation, declining birthrates, rising inequality—these are the defining challenges of our modern economic reality, and we all want someone to blame for them, naturally. That way we can sort out how to fix these issues, or so the thinking sometimes goes.

Concentrated wealth driven by rapacious, greedy Wall Street bankers and their bought politicians have ground the economy to a halt, some would convincingly argue (including some economists subscribing to the secular-stagnation hypothesis, seeing far too much money stuck in the hands of wealthy individuals who by and large don't spend it). Others argue that a long-running war on labor unions and workers in general has fueled gross income inequality or that corporations used their political power to successfully pass regressive tax policies and other laws designed to make the rich richer at the expense of everyone else. And many still argue that Americans are having far fewer children than in past decades because of their high student loan burdens and the failure of Congress to mandate paid maternity and paternity leave, among other economic reasons. These are all imminently more satisfying explanations for these conundrums because they give us someone to blame, someone or some group to point at and say, "This is all your fault." That way, some believe, we can fix matters by removing these nefarious actors from positions of power and reversing what they have done to bring us here.

But again, unfortunately, these are not the root causes of our most vexing economic challenges, as I hope I've convinced you, based on the theories of the great alternative thinkers introduced to you in these pages. We've been through all this before, but allow me to cover this ground again for a brief moment.

To reiterate somewhat, there are European countries that have free college tuition and even mandatory paternity leave, and those same nations' birthrates are generally lower than America's. A solid social safety net or social contract does not seem to correlate well with numbers of births per couple. Meanwhile, wealthy campaign contributors have indeed co-opted their favorite politicians to lower their taxes and eliminate regulations and barriers to their investments, but they did so in large part because they assumed that this would all lead to faster economic growth, thus faster returns. That's certainly why the mainstream economists advocated in favor of such policies in the first place—they were operating under the same flawed assumptions. In addition, the slow and steady economic growth funk befalling us now very much threatens the influence peddlers' wealth alongside the economists' credibility (especially in terms of the ongoing global political backlash). And progressive thinkers themselves have, for the most part, been vocal champions of open borders, freer migration, freer trade, and globalization in their own right and probably won't be advocating for a reversal of these trends anytime soon, even though the rise in global inequality can be linked to all these developments, since these forces end up "better mixing" the global economy and thus accelerated the economy's natural tendency toward maximum entropy. Regressive taxation could be playing a role, I suppose, but probably a relatively minor or supporting one.

All this is not to entirely dismiss or even discount the more popular judgments concerning what's driving large macroeconomic challenges in any way. Americans really do feel the pinch from college loan debts and weak family-friendly laws and policies, and this no doubt influences their family-planning decisions. Politicians really do obey the rich first and the rest last, as studies have shown, and this probably does contribute to gross economic inequality to a great extent. And gross domestic product growth really does slow when innovation and technological progress are weak or relatively anemic. But these are perhaps best understood as symptoms or tangentially related factors. In the first three chapters, this book offers density dependence, EROI, and entropy as "core" explanations for these broader trends, as the central points from which all other factors originate or emanate. They are the starting points from which everything else can be explained or more thoroughly understood. These are natural forces now having their way with us.

But the real reason why some progressive-minded individuals will probably raise loud objections to what I have laid out in these pages is, I suspect, the utter lack of practical solutions or cures presented or proposed. I acknowledge that it's an unusual approach for a text of this sort. Normally an author would lay out an argument for how things work in detail—the world according to him or her, more or less—and then spend many pages explaining how things should be and what we all should do to get there. Define the problem, present a new and insightful explanation for it, and then propose a novel and unexpected resolution—that's the typical formula for a book of this sort, right? But I'm not your typical nonfiction author, I'm afraid, and this is not that kind of book. Many popular prognosticators would be appalled, but that's just the way it is.

Yes, the theories shared in these pages, developed by brilliant yet far-too-often-ignored scientists and academics, include within them no concrete solutions to the global economic challenges they break down for us in exquisite clarity. As far as the birthrate issue is concerned, there is no fix, as I've already argued. Density dependence works kind of like gravity, and all you can really do is let it take its course and roll with the changes as best as you can. Entropy will always be a part of our universe; we can arrest it temporarily for sure, by applying energy to reorder things as we like, but the path toward disorder will eventually set in again, usually immediately after we take our eyes off the ball, as will inevitably happen. Our energy and growth conundrums may have rather good, workable solutions, but none that are immediately on the horizon as far as I and many others can see. If there is an energy transition under way, it's from energy-dense, relatively higher-EROI fossil fuels to less energy-dense, lower-EROI renewable-energy technologies. My hope for the eventual success of the nuclear fusion experiments under way is little more than that: mere hope, not certainty or expectation. And there may be something inevitable about declining EROI as well, given that no system, organism, or colony has yet found a loophole out of it.

But for those who hate or reject the arguments laid out in detail in these pages, for reasons having to do with a progressive appeal for a brighter future, I can offer a small bit of solace.

It should be noted that the women and men of science championed in these pages are hardly an omniscient lot, and it has long been apparent to me that they don't have all the answers either. For instance, in my conversations with the self-described biophysical economists, I found that many of them, if not most of them, are fairly unfamiliar with the concept of density dependence, a factor in animal biology and physiology at least as important as caloric intake and energy return on investment. Or if they are knowledgeable about the concept, they at least haven't recognized its critical role in human population dynamics and economic growth.

And near as I can tell, the very small number of scholars who earlier landed on rising population density as the explanation for declining human birthrates have since abandoned the very explanation themselves, probably because it didn't catch on with their peers and was generally dismissed by the academic literature. Hence the ongoing and overwhelming emphasis on multiple and sometimes even contradictory socioeconomic factors to explain the phenomenon instead. Density dependence is the best, most brilliant, and correct explanation, but it remains an unpopular one, so academics by and large ignore it, even those scholars who proposed it in the ivory tower journals in the first place. In an e-mail, Lutz confessed to me that he dropped this line of inquiry due to peer pressure. (His colleagues found the idea distasteful and told him to stop, so he did.)

Furthermore, Victor Yakovenko and other scholars zeroing in on findings that entropy is the main factor driving greater global economic inequality have also written of their beliefs that a broader uptake of renewable-energy technologies like wind and solar power in the world's energy mix may eventually alleviate this very economic inequality. Yet the low EROI ratings earned by those technologies would lead others to seriously question this argument.

Put together, the density dependence, EROI, and economic entropy hypotheses are powerful conceptual tools for comprehending the greatest and most dynamic macroeconomic trends moving our human economy/ecology today, but they have only been unified within this text as far as I am aware. They remain balkanized, and thus marginalized, in academic circles and discussions. Their practical impacts on scientific inquiry and policymaking are pretty much negligible, and their influence in the professional field of economics is nonexistent.

I am also obviously no expert on ants. If you picked up this volume to learn more about the ant universe, then I'm afraid you've just read the wrong book, and I apologize.

We should definitely leave it to the entomologists to teach us about the amazing complexity and sophistication of ant colonies. My point was to draw simple parallels between their world and ours, to demonstrate for you how certain patterns in nature found on Earth can play out across two radically different orders of animal with one thing in common: extreme social organization. We imagine our cities, agriculture, and complex civilization to have emerged over long periods and according to a process of our choosing. This book invites you to imagine that things may have worked rather differently, that our arrival at this point in our evolution may have been entirely inevitable and not a consequence of any conscious debate, deliberation, or decision making by our leaders. The ants, after all, build cities, farms, ranches, gardens, highways, and landfills while arranging themselves into workers, soldiers, hunters, foragers, caregivers, undertakers, and waste-disposal crews, all without thinking things through first. I believe much of the same can be said about our human civilization as well: we didn't choose to build cities, grow crops, raise livestock, and build a complex hierarchical social order; we were compelled to do all these things by the forces of nature and our surrounding environment. That's my point, and the various interesting examples ant experiences provide go far in making the case, I believe.

But yes, this book is probably flawed in many ways. There is plenty of conjecture on offer in these pages to accompany the facts presented alongside. Thus, many more critical and even cynical readers may deem the following quote attributed to Mark Twain a rather fitting summation of the arguments put forth herein: "There is something fascinating about science. One gets such wholesale returns of conjecture out of such a trifling investment of fact."[3]

This is fair criticism. And yet, I stand by my broader premise.

The theme of this book is how general rules governing the dynamics of animal ecosystems and the natural environment play a big role in governing our own human economy/ecology as well, and I set out to demonstrate how by leaning on some of the more fascinating ideas and theories put forth by my heroes, the contrarian natural scientists, ideas that the mainstream leaders of economics and policy debates are unfortunately ignoring. It all amounts to a trifling investment of facts, I suppose, but they are important facts that deserve closer consideration.

And though my arguments may seem like mere "wholesale returns of conjecture," the facts that nature presents to us do offer powerful insights into what we as a species are and how we, collectively, arrived at this point in space and time. It turns out that the lessons informing humanity's own social evolution have been on display all around us this entire time, reflected in the mirror back at us in the form of the animal and insect world, requiring of us only a willingness to pay attention, to see what general rules or guidelines seem to apply to nature and then imagine how they almost certainly apply to us as well.

The lessons are worth repeating. Nature teaches that clumping or clustering, the tendency of animals to group together, is a natural law and tendency that affects nearly all species, plants and animals, and that this is not unique to humans. We humans, along with ants and termites, merely take it to extreme levels. And by looking to the ants for lessons, we can see how this sort of extreme degree of clumping can influence social animal evolution. For instance, nature appears to dictate that higher orders of complexity in resource acquisition and allocation leads next to higher orders of complexity in social organization, not the other way around. And again, this is not unique to humans, though many of us may have long assumed that it was. We can also find clues to how behavior and labor roles in social animals can change and how these developments are emergent phenomena, not necessarily predetermined in any concrete way, just as none of us is genetically preordained for the roles that we all play in our own societies, though our interactions with each other heavily influence the eventual roles or statuses we end up in.

Nature can also prove to be just as obsessive about the critical importance of waste management as we are. Garbage and waste hauling are messy, time-consuming, and expensive but critical to mass-colony organism survival. Note how ants can and do exert enormous amounts of time and energy to merely ridding themselves of waste because of the grave threat that it poses to their colonies and crops. And we can begin to see why direct reuse or recycling of these wastes is so rare in animal and human cities alike: the cost-benefit analysis rarely tips in favor of waste recycling and reuse. But occasionally it can, especially in an environment where acquiring and distributing resources and energy the standard ways sometimes becomes more difficult or expensive. All of these facts are laid out by the natural order of things and are not man-made realities; they thus potentially offer further clarity for our own human conundrums.

Those were some of my own insights into how the ant world could help illuminate the development of our own. Far more pertinent to this discussion, of course, are the economic ideas laid out in the first three chapters: density dependence's drag on birthrates, energy return on investment's drag on economic growth, and the natural tendency of economies to seek entropy, not equilibrium, leading to higher orders of inequality and a drag on greater social justice.

Though these ideas emerged from naturalists and the hard, physical scientists, they yield great power in better illuminating human social dynamics as well. These overarching neoeconomic concepts are advanced mainly by noneconomists, but the economists should pay attention to them, because they offer us focused explanations with great predictive value. They are simplified models derived from the natural world employed with great effect to explain highly complex macroeconomic human phenomena. And these ideas happen to work, unlike many other popular economic concepts that have failed us time and time again.

Above all, in this book I have sought to demonstrate convincingly that many talented scientists and naturalists happen to know quite a lot about economics, even if they don't immediately recognize this fact themselves.

And I believe I've further underscored how most economists, probably the vast majority in fact, know next to nothing about the natural world and how it pertains to their own chosen discipline. Why this is, I don't know, but it could stem from the fact that the field of economics, as it is mainly practiced, is among the least curious professions and disciplines in academia today. By this, I mean that I can find little evidence of a mass movement toward embracing more interdisciplinary studies among the world's university economics departments, even as other fields of study have entirely embraced interdisciplinarity in a very big way.

Astronomers will frequently look to particle physicists for lessons on how the cosmos operates, and vice versa. Biologists will pick the brains of chemists and even geologists, and geologists often have full appreciation for the biological processes involved in forming a host of minerals that matter greatly to their research. Political scientists frequently consult mathematicians, sociologists will read through medical journals, and environmental scientists will borrow bits and pieces from virtually all hard and social sciences in their own investigations. This very spirit of interdisciplinarity drove physicists, chemists, biologists, and even mechanical engineers to dabble a bit in economics in the first place. Yet the economists seem to be, for the most part, keeping themselves deliberately walled off from their intellectual peers and academic cohorts residing in other parts of the ivory tower. I'm not the first to point this out, and I certainly won't be the last.

Remember when Yakovenko lamented about how the economists working on the very same campus as he routinely shun and ignore him? He doesn't seem to mind though, probably because he recognizes that this is their loss and not his. And it is a grave loss for them. If they would only open their eyes to some fascinating natural rules and patterns and the lessons available from animal ecosystems, then they might solve many of the mysteries that continually elude them.

That may include their greatest mystery: why economics has proven such a failure in recent years. Yakovenko offers one possible explanation, and he offers it free of charge. Economists seek out equilibrium in their modeling, thus their fetish for equilibrium distributions in a host of theorizing and analysis. However, as physicists would point out, in nature (i.e., reality) equilibrium is largely an illusion. The real natural tendency is toward entropy, meaning maximum inequality and disorder and eventually stagnation, because unequal and stagnant systems are stable, whereas systems in equilibrium are unstable. This natural law thus strongly implies that the pursuit of deregulation, free market economics, and free trade is, in reality, a perfect recipe for greater inequality and stagnation in the long run and not for the unstoppable growth and harmony and equilibrium (or equity) that the economists incorrectly assume will result. Of course, this describes precisely our current economic/ecological dilemma, if you think about it. But economists are largely blind to this possibility because they are far too often blind to the natural order of things. Make no mistake, you and I and they and everyone else on this planet are products of a natural world, not the invented mathematical worlds that many of the economic schools prefer.

The rules of nature strongly imply that achieving a more equitable outcome, or an equilibrium distribution in any real-world economy, requires much more active intervention, not less, including possibly more regulation, more taxation, higher levels of directed and targeted government spending, and even direct government intervention or leadership in industrial policy. All horrifying prospects for many economists, I'm sure, but the alternative appears to be higher levels of inequality and slower or even stalled growth, at least according to the rules of nature as outlined above. After all, according to principles laid out in physics, the only way to arrest entropy in a system is to apply energy to reorder it; otherwise the system, left to itself, will move toward greater disorder. This implies that a "free market" economy must never be left truly free unless the economists' ultimate goal is global economic stagnation. When an economy and any and all such randomized systems are allowed to order themselves, they will do so along a path toward greater entropy, not greater equilibrium. That's a powerful concept from the natural world that is absolutely relevant to our economy/ecology, I strongly believe.

Physicists have also long understood that what works at smaller scales does not necessarily work at much larger scales, offering further insights into how economists might be getting things so wrong all the time. Perhaps trade relations indeed follow along the lines envisioned long ago by David Ricardo when only two actors are involved, but not when you have two hundred, and certainly not with two thousand or two million or two billion. Equal agents with equal endowments do not forever remain equal when interacting in very large numbers or at very large scales. Instead, a sort of phase transition occurs at larger-scale interactions, and supposedly certain relations or certain outcomes become probabilistic and randomized. Understanding this fact, one may conclude that relying on theories based on simplified smaller-scale models poses great risk if one simply extrapolates their implications to much larger scales. Extrapolation from smaller scales to larger scales doesn't necessarily turn out alright in investigations of the natural world, and economies are driven by people, and people are products of nature. So why should we expect any different outcome with economics?

As you've read in these pages, there are further lessons on offer from nature and the animal kingdom that can better inform our understanding of our own human economy/ecology. Chief among these: the central importance of energy, which is at the heart of everything.

Our human economy/ecology is much more than supply and demand, capital and labor, or renters and rentiers. It describes the myriad ways in which we acquire and consume energy to transform matter and other resources into materials of value and importance to us, as well as how we distribute these materials and this energy throughout our society, efficiently or no. This is how it works for other organisms, including the ants. The strength and fidelity with which this process occurs or is accomplished is known to naturalists as the energy return on investment. If average societal EROI is in decline, then does it not stand to reason that this is having a very great impact on the human economy/ecology, the same as declining EROI would with organism health and fitness? Here again, many economists are often too readily dismissive of this possibility because they regard energy as just another input that results in economic outputs, or as another tradeable product like milk or cheese or textiles or refrigerators. It's an overly simplistic way to look at the world: none of those other things could exist without energy, but energy can easily be realized without any of that other stuff. The human economy itself wouldn't exist without energy; the same goes for antelope herds, flocks of Canada geese, prairie dog colonies, and anthills.

The animal kingdom also teaches us about what happens to human biology and reproduction at higher population densities, lessons unknown to economists who regard growth as sacrosanct and eternal and crowded cities as the engines of this everlasting economic growth. As you read earlier, conditions of overcrowding drive both physical and behavioral changes in mammals that have a very real, material impact on their economy/ecology. Density dependence explains precisely the collapse in fertility and birth rates globally better than any other socioeconomic explanations frequently put on offer so far. Economists are again ignorant of this, and many earnestly believe that only they have answers regarding why this is happening and how to fix it. Would-be parents are short on cash; thus "baby bonus" payments are the impetus needed to encourage them to breed, many suggest. Or we need stronger paternity- and maternity-leave laws, others propose, in order to give aspiring parents more confidence that their careers will not be interrupted or derailed by bringing new life into the world. But none of that will actually work, and the reasons these policies fail will likely continue to elude the economists because, again, too many of them will continue to ignore the natural order of things. They will remain stuck in seeing declining fertility and birth rates as the result of relatively simple, if stubborn, economic or social problems and not as a natural consequence of high animal population densities. But the core explanation behind the ongoing global baby bust is overcrowding and nothing else. I know I'm being awfully repetitive here, but it's worth it to further hammer this point home in case it hasn't fully gotten across to some of the more skeptical of you reading these words now.

Again, I am no scientist, just a scholar and observer attempting to better understand our human reality by focusing on the preexisting laws and rules governing the animal kingdom and the natural world, the same natural world that gave rise to all of us. Many will not agree with my approach and will find different ways to poke holes in my arguments, I'm sure, and that's fine. But what's the bigger crime here: a lay observer's clumsy leaning on science and scientists in an effort to better understand the laws of nature and how they may apply to our very own economy/ ecology or the economists' approach of continually ignoring these long-established natural laws and entirely inventing their own instead? And it is these invented laws, and not the natural ones, that dominate and command their assumptions and underpin their most heavily relied-upon economic models and theories. Trouble is, these standard approaches appear to be failing the mainstream economists as of late, having left far too many of them blind to potential outcomes that they didn't anticipate or otherwise downplayed—that is, until after all these unexpected consequences blew up in everyone's face.

There's a lesson in store for mainstream economists here. Their models may no longer fit with established reality, if they ever did. Therefore, the correct path moving forward is not to double down on the standard approaches or to push for policies designed to alter the prevailing reality or bend it in some way to fit the standard modeling. The correct response, I think you now know, is to change the fundamental assumptions behind these models and, ultimately, underpinning most of modern economics.

My conclusions or reasonings may sound a tad harsh, but I actually consider myself to be among the more generous critics of the economics profession. There are plenty of harsher critics out there whose anti-economist views have been published and widely disseminated.

You see, I'm a big believer in the concept that it's possible for someone to become so brilliant as to accidentally render him- or herself an idiot. This likely explains a lot of the economists' missteps. They are frequently brilliant, for sure, but this brilliance has far too often led them into arrogance, and in turn ignorance, as their skills in mathematics and econometrics have caused far too many of them to miss the forest for the trees. Their intellectual prowess has resulted in their becoming far too dismissive of critics and contrarian points of view, confident as they are that they have all the answers and those who object to their explanations and insights are intellectual lightweights with no business advising or criticizing them.[4] Many of them have become so brilliant that they've accidentally rendered themselves idiots. Perhaps more recent geopolitical and geoeconomic events will lead these geniuses to humble themselves a bit and to be more thoughtful and willing to see the world through different lenses and different disciplines. But I'm not holding my breath.

Other harsher critics of economists would probably vehemently disagree with this take and say that reform on their part is an impossibility and that wishing for any change of heart to sweep over the modern economics profession is a complete waste of time. For example, Thomas Frank is a renowned journalist and author of several books critical of much U.S. government policy that has relied far too heavily on economists for guidance, in his view. Here's how he described the economics profession during an October 2017 interview on the online news and opinion program *The Young Turks*: "It's like a cartel invented to defend error and to protect folly," Frank told the host. "People who they disagree with are sort of shunted aside to second-string colleges and are never in the advice-giving big leagues, ever. That is the nature of that profession."

Frank seems to believe that the economics profession isn't about to change anytime soon. He may be correct, and the main ideas put forth in this text probably won't percolate throughout the broader economics community to any serious degree in the near future. And most likely, I will never find myself in the "advice-giving big leagues" either. But allow me a moment to pass on some helpful tips anyway to any policymakers who may be reading these words now.

Nature holds the clues to developing fundamental, simplified, yet large-scale models that can explain a host of macroeconomic phenomena, and if these phenomena are indeed natural in origin, then the best policy approach may be to simply let these trends take their natural course, as that's what's going to happen anyway. Thus, we are free to ignore the "empty planet" hyperbole loading arguments surrounding falling fertility and birth rates, for example. Many experts will disagree, of course, but their reasons for doing so are often highly nonsensical, and the "solutions" they propose are impractical and unworkable.

Take this argument, for instance: "When population growth is negative, both endogenous and semi-endogenous growth models produce what we call an *Empty Planet* result: knowledge and living standards stagnate for a population that gradually vanishes. . . . In contrast, if the economy switches to the optimal allocation soon enough, it can converge to a balanced growth path with sustained exponential growth: an ever-increasing population benefits from ever-rising living standards."[5] Either the individual who wrote these words was possessed by a host of popular yet entirely baseless assumptions about the decline in global birthrates, or he was high.

For starters, population decline does not mean a population is doomed to vanish entirely. Reporters or pundits, like the author quoted above, who are all too fond of casually musing about the inevitable extinction of South Koreans, Italians, or other nationalities should have their heads examined, in my humble opinion. These are already massively overcrowded nations that would have found it rather difficult to become even more crowded and populated, even had they the will to do so. Since they don't, what we can expect is (likely inevitable) population declines in these nations until an optimal number is reached whereby the existing populations are not overstressed to the point that average fertility is dragged lower. Once these nations escape stress, they will be rid of the natural forces of density dependence and free to be fruitful and multiple once again. But that's not going to happen anytime soon.

Furthermore, "sustained exponential growth" and "ever-increasing population" are physical impossibilities in finite systems, and our species most definitely exists in a finite system. Anyone who doesn't understand this basic fact has no business advising governments, business leaders, policymakers, or anyone else about what optimal birthrates and population numbers ought to be.

Ask this question of any person on any crowded urban street, be it in Manhattan in New York City, Ginza in Tokyo, or Piccadilly Circus in London: "Hey you, what do you think, do we need more people on this planet, or fewer?" What do you honestly expect their answer would be? I think I already know what most people would say in response, and they wouldn't give a damn that most professional economists or the individual quoted above disagreed with them. The fact is that global birthrate decline seems to be of natural origin, though it has obvious socioeconomic implications, and there is no reversing it so long as human population density continues to increase via ongoing population clumping, the phenomenon otherwise known as urbanization. And we shouldn't wish to reverse or combat it anyway: if it's occurring, this is because it *must* occur per the natural forces at play driving this trend.

Falling birthrates and population declines could even prove to be enormously beneficial, contrary to popular belief. I sincerely mean this. In fact, let's have the courage to think a little differently for a change. Instead of assuming only bad things will come from the ongoing fall in global average human fertility, I advise us to imagine for a moment some of the unexpected positive outcomes that could arise instead. We may find ourselves pleasantly surprised, and not only in environmental terms. If crowding has led to higher living costs, fiercer competition for decent employment, stagnant wages, greater anxiety about the future, and, in reality, a general decline in standards of living—all the stressors behind the fall in fertility—then perhaps a future path toward societal "decrowding" will deliver to us the exact opposite outcomes. It isn't beyond the realm of possibility. In fact, a few brave out-of-the-box-thinking economists have already delved into the question of what might happen in an environment of declining population and have discovered that it is entirely possible that per capita output and welfare can actually increase as a nation's population contracts.[6] I'm seeing some evidence in support of this idea already arising here in Japan: in my "shrinking" city, several new businesses have opened up recently, including many owned and operated by young people who probably would have faced far greater difficulties investing in still-expanding urban centers like Tokyo. Yes, there will be challenges to face in terms of meeting pension and retirement fund obligations and health-care expenses, but these aren't necessarily insurmountable problems. Allow me to humbly propose one possible way forward: let's stop thinking of the elderly as useless parasites and instead imagine ways that they can be empowered to contribute to the greater social welfare.

There is also, of course, immigration, an effective, if very much temporary and imperfect, means of responding to the trend. It is certainly true that a government can thwart its population's refusal to procreate by importing hundreds of thousands or even millions of new individuals to fill the gap. It's one policy option to pursue, so long as its proponents fully understand its limitations and what higher immigration will not deliver: it will not raise wages, it will not make housing more affordable, it won't make cities less crowded, and it definitely will not raise a society's average birthrate over the long term. But this is a conversation I'll leave for others to pursue.

Of course, policymakers are free to continue consulting their trusted economists. How could I ever stop them? But here is some further advice: invite some alternative and contrarian voices and viewpoints into the debate mix once in a while, and don't immediately dismiss these fresh and different ideas and perspectives, even though the economists you consult with will surely demand that you do so. Then challenge your economist consultants to also pause and take seriously what these contrarian thinkers have to say, if only for a moment. At a bare minimum, they may (hopefully) find compelling the notion that although they seek out equilibrium in their econometric thinking, nature always seeks out entropy, so it might be of some utility to alter their thinking to align more closely with this fundamental reality instead.

They may also find that they should forever toss out the notion that economic growth can and should last forever. Just as there is no real equilibrium state in animal ecosystems, there is no eternal growth found in nature either. What we find instead is a constant ebb and flow, a cycle of expansion, stagnation, and contraction. And, for life, eventual extinction. It may look like equilibrium initially, but up close we find that it is anything but. Again, is there any good reason to assume that our economy/ecology works any differently? I think not. The economic models may call for continuous eternal growth, but in nature this is an impossibility, and natural reality could now be reflecting this very fact for us in

several ways that the elite economists never initially anticipated. To repeat: the key may be not to manipulate reality or to seek loopholes in the
rules of nature but rather to change some of our fundamental assumptions. In other words, we need the courage to change our thinking about
what actually constitutes some of the more basic fundamentals of macroeconomics and how the natural world may actually limit or restrict economic growth, whether we like it or not. Constraints are very real in
nature; they must be in economic realities as well.

I have some additional advice below, so bear with me a bit longer.

Energy probably needs to be placed front and center in how we com-
prehend our human economy/ecology, meaning we should make an attempt to more thoroughly understand how our economy/ecology functions and operates, or how it evolves and changes, in terms of how our
primary energy sources and their relative abundance evolve and change
in tandem. After all, at its heart the global economy involves making and
then trading stuff and services, and it takes energy to make all this stuff
and to deliver all these services in the first place. From this point of view
energy seems rather important, and our human economy/ecology's overall energy return on investment must be rather pertinent in determining
just how much output per person our various economic activities can
ultimately derive. Until we more fully appreciate this fact poor, Larry
Summers may find himself stuck describing a problem without fully
understanding it, forever assuming that secular stagnation has everything
to do with psychology and technology and nothing to do with energy.

Again, with this writing I've attempted to demonstrate convincingly
how natural rules common to animal ecosystems might help to explain
very large economic trend lines and how the contrarian thinkers of our
time can reveal this reality to us. They have valuable lessons to impart, if
only we would take the time to listen or give them a larger platform. A
broader diversity of viewpoints can only be healthy for any vigorous
debate, be it in economics or any other discipline.

These alternative economic thinkers, stepping out of the comfort
zones of their hard science backgrounds, are offering compelling explanations for why physical, natural reality seems to crashing up against our
conventional, everyday economic thinking—and winning. Expanded global trade and globalization doesn't seem to be boosting average global
GDP growth rates; rather, the global economy appears to be slowing
down precipitously instead. Free markets and free trade don't seem to be

alleviating inequality either, only exacerbating it. And ongoing urbanization is not delivering strong growth and happy families but rather economic and population stagnation.

Natural laws could have easily predicted these outcomes long ago,
whereas the economists never did. Thus, it may be high time to invite the
naturalists to the table as these discussions unfold. What do we have to
lose? I can see no downside to this approach, only an upside.

It may not seem like it, but there is much hope and optimism bleeding
out of these pages. I hope that we will embark on the process of expanding our knowledge and understanding of the world that we inhabit and
how the environment and our emergence from it determine our collective
path. Knowledge is power, and this power will eventually set us free. Our
fates aren't necessarily written in stone, after all. We are still fully capable of influencing the trajectory that nature has set us out on. The first
step is understanding and appreciating the forces working upon us; then
we may find a means of working within these natural confines to build a
world that eventually ensures maximum peace and happiness for the
maximum number of individuals possible. It's a goal worth pursuing.

NOTES

1. DENSITY DEPENDENCE

1. Anna Louie Sussman, "The End of Babies," *New York Times*, November 16, 2019.

2. Larry Elliot, "A Birthrate Crisis Would Require a Whole New Mindset on Growth," *Guardian*, March 31, 2019.

3. David Stanway, "China Lawmakers Urge Freeing Up Family Planning as Birth Rates Plunge," Reuters, March 12, 2019.

4. William Reville, "Earth's Population May Start to Fall from 2040. Does It Matter?," *Irish Times*, March 7, 2019.

5. Sanjeev Sanyal, "The Wide Angle: The End of Population Growth," *The Wide Angle* (Deutsche Bank), May 13, 2011.

6. Park Hyong-ki, "Economy Faces Demographic Shock amid Low Birth Rate," *Korea Times*, March 28, 2019.

7. Jacqueline Howard, "U.S. Fertility Rate Is below Level Needed to Replace Population, Study Says," *CNN*, January 10, 2019.

8. Jason Horowitz, "Italy's Right Links Low Birthrate to Fight against Abortion and Migration," *New York Times*, March 27, 2019.

9. Ariana Eunjung Cha, "As U.S. Fertility Rates Collapse, Finger-Pointing and Blame Follow," *Washington Post*, October 19, 2018.

10. I've come across self-appointed blogging "experts" blasting media coverage of this topic because reporters tend to conflate birthrates with fertility rates, and vice versa. Such criticism is misplaced and should be dismissed. The birthrate and the fertility rate are, at the end of the day, calculating the very same thing: how many babies are born. It's the difference between measuring the temperature in Fahrenheit or Celsius.

11. Ned Rozell, "What Wiped Out St. Matthew Island's Reindeer?" *Anchorage Daily News*, January 16, 2010.

12. Greg Yarrow, "The Basics of Population Dynamics," Clemson Extension, Fact Sheet No. 29, Forestry and Natural Resources, May 2009.

13. Yarrow, "The Basics of Population Dynamics."

14. Ben C. Scheele et al., "Amphibian Fungal Panzootic Causes Catastrophic and Ongoing Loss of Biodiversity," *Science* 363, no. 6434 (March 29, 2019).

15. Colin Carpenter, "The Ups and Downs, Hows and Whys of Wildlife Populations," *West Virginia Wildlife* (Fall 2013).

16. Doug P. Armstrong et al., "Density-Dependent Population Growth in a Reintroduced Population of North Island Saddlebacks," *Journal of Animal Ecology* 74, no. 1 (2005).

17. Christophe Bonenfant et al., "Empirical Evidence of Density-Dependence in Populations of Large Herbivores," *Advances in Ecological Research* 41 (2009).

18. Robert K. Swihart et al., "Nutritional Condition and Fertility of White-Tailed Deer (*Odocoileus virginianus*) from Areas with Contrasting Histories of Hunting," *Canadian Journal of Zoology* 76, no. 10 (1998).

19. V. Bretagnolle et al., "Density Dependence in a Recovering Osprey Population: Demographic and Behavioral Processes," *Journal of Animal Ecology* 77, no. 5 (2008).

20. Andrew G. Clark and Marcus W. Feldman, "Density-Dependent Fertility Selection in Experimental Populations of *Drosophila melanogaster*," *Genetics* 98, no. 4 (August 1981).

21. As of this writing the spread of the novel coronavirus from China to the wider world is seriously challenging this assumption, while underscoring the exacerbation of our susceptibility to limiting factors and decimating factors as a result of excessive population crowding.

22. Wolfgang Lutz and Ren Qiang, "Determinants of Human Population Growth," *Philosophical Transactions of the Royal Society of London* 357, no. 1425 (2002).

23. It should be noted that high levels of urbanization and urban population density are definitely key variables, but the effect of crowding in cities appears to reverberate well beyond city limits. Reports suggest, for instance, that birthrates in the United States are falling in rural areas as well as urban centers.

24. Wolfgang Lutz, Maria Rita Testa, and Dustin J. Penn, "Population Density Is a Key Factor in Declining Human Fertility," *Population and Environment* 28 (November 2006).

25. Lutz, Testa, and Penn, "Population Density."

26. Mary Regina Boland, "A Model Investigating Environmental Factors That Play a Role in Female Fecundity or Birth Rate," *PLOS ONE*, November 27, 2018.

27. David de la Croix and Paula E. Gobbi, "Population Density, Fertility, and Demographic Convergence in Developing Countries," *Journal of Development Economics* 127 (2017).

28. Teresa Janevic et al., "Effects of Work and Life Stress on Semen Quality," *Fertility and Sterility* 102, no. 2 (August 2014).

29. Yes, Canada is massive, yet one-third of all Canadians live in just three tightly packed urban districts: Toronto, Montreal, and Vancouver.

30. For one possible example, Canada has higher rates of immigration per capita than the United States and a higher proportion of foreign-born in the population. Population growth in Canada is now driven primarily by immigration since Canadian fertility is so low: 1.6, below the United States and Finland and slightly above Germany. Population increases there are slowing, according to the scenarios laid out by Statistics Canada and the United Nations.

31. Lyman Stone, "The Global Fertility Crisis," *National Review*, January 9, 2020.

32. I'm aware of government and media reports showing life expectancy already declining in the United States, but I don't think that density dependence is the cause. Rather, the likeliest explanation for this depressing trend can be found in Chapter 3 of this book.

2. EROI

1. William A. Strauss and Kelley Sarussi, "Economic Growth to Decelerate in 2019 and Then Ease Further in 2020 as Auto Sales Downshift," *Chicago Fed Letter*, No. 417 (2019).

2. International Monetary Fund, World Economic Outlook Report, October 15, 2019.

3. United Nations Department of Economic and Social Affairs (UNDESA), "World Economic Situation and Prospects 2020," January 2020.

4. UNDESA economists should be taken seriously. Unlike the supposed geniuses who run the World Bank, IMF, Federal Reserve Bank, and other institutions, UNDESA stands alone in actually warning of the U.S. housing bubble of the early 2000s and predicting global economic havoc should it pop, which is exactly what happened.

5. "Credit Conditions—Global, 2020 Outlook," Moody's Investors Service, November 19, 2019.

6. Richard Baldwin and Coen Teulings, eds., "Introduction," in *Secular Stagnation: Facts, Causes, and Cures*, ed. Baldwin and Teulings (London: CEPR Press, 2014).

7. Baldwin and Teulings, "Introduction," 32.

8. Baldwin and Teulings, "Introduction," 33.

9. Some scholars use the term "energy return on energy invested" (ERoEI), but it's the same thing.

10. Jessica G. Lambert et al., "Energy, EROI and Quality of Life," *Energy Policy* 64 (2014).

11. Lambert et al., "Energy."

12. Lambert et al., "Energy."

13. Jessica G. Lambert, Charles A. S. Hall, and Stephen Balogh, "EROI of Global Energy Resources: Status, Trends, and Social Implications," commissioned by the United Kingdom Department for International Development, October 2013.

14. Lambert et al., "Energy."

15. Adam Lampert, "Over-exploitation of Natural Resources Is Followed by Inevitable Declines in Economic Growth and Discount Rate," *Nature Communications* 10, art. no. 1419 (2019).

16. Bill Chappell, "U.S. National Debt Hits Record $22 Trillion," *NPR*, February 13, 2019.

17. M. Ayhan Kose et al., *Global Waves of Debt: Causes and Consequences* (Washington, DC: World Bank, 2019).

18. Pumped hydro storage proposes using underutilized or unused nighttime wind power generation to move water uphill via pumps. The water would then be released, allowing gravity to pull it downhill and through turbines to generate new electricity in the daytime, thereby sort of "storing" the wind energy for later use. It's a great idea, but its existing deployment is very limited, and thus the technology currently doesn't make any significant dent on energy markets.

19. D. Weissbach et al., "Energy Intensities, EROIs (Energy Returned on Invested), and Energy Payback Times of Electricity Generating Power Plants," *Energy* 52 (2013).

20. Weissbach, "Energy Intensities."

21. Paavo Järvensivu et al., "Governance of Economic Transition," Global Sustainable Development Report 2019, drafted by the group of independent scientists, August 14, 2018.

22. Järvensivu et al., "Governance of Economic Transition."

23. You would be amazed at the volume of shipping fuel and diesel it takes to install just one offshore wind-power facility.

24. "25th ITER Council: All Efforts Converging toward the Start of Machine Assembly," ITER press release, November 21, 2019.

3. ENTROPY AND INEQUALITY

1. Here I refer to the election victories of such figures as Donald Trump in the United States, Boris Johnson in the United Kingdom, and Jair Bolsonaro in Brazil. Such outcomes have been deemed more than once by critics as "threats to democracy," which is ironic given that they are the results of democratic elections.

2. This according to research and data compiled by the International Federation of Robotics.

3. Thomas Piketty, *Capital in the Twenty-First Century* (Cambridge, MA: Harvard University Press, 2014).

4. Piketty, *Capital*, 746.

5. OECD, *Under Pressure: The Squeezed Middle Class* (Paris: OECD Publishing, 2019).

6. Elaine Kurtenbach, "Report: Rich Will Still Get Richer unless Policies Change," Associated Press, Tokyo, December 15, 2017.

7. Ali Alichi, Kory Kantenga, and Juan Solé, "Income Polarization in the United States," IMF Working Paper, International Monetary Fund, 2016.

8. "Inequality Illusions: Why Wealth and Income Gaps Are Not What They Appear," *The Economist*, November 30, 2019.

9. Republican Party nominee Donald Trump's victory over Democratic Party nominee Hillary Clinton in November 2016, of course.

10. Robert E. Scott, "The High Cost of the China-WTO Deal," Economic Policy Institute Issue Brief #137, February 1, 2000.

11. Scott, "The High Cost."

12. David H. Autor, David Dorn, and Gordon H. Hanson, "The China Shock: Learning from Labor Market Adjustment to Large Changes in Trade," National Bureau of Economic Research, Working Paper No. 21906, January 2016.

13. Named for the Austrian physicist Ludwig Boltzmann and the American scientist Josiah Willard Gibbs, who together developed mathematical models that further explained the laws of thermodynamics and how they operate in complex systems.

14. V. Yakovenko and A. Dragulescu, "Statistical Mechanics of Money," *European Physical Journal* B 17 (2000).

15. Ricardo's famous idea holds that two nations trading with one another can maximize welfare through one nation specializing in products that it can produce more cost-effectively or efficiently than its trading partner, and vice-versa, thereby having these two hypothetical countries trading along lines of "comparative advantage."

16. Victor M. Yakovenko, "Applications of Statistical Mechanics to Economics: Entropic Origin of the Probability Distributions of Money, Income, and Energy Consumption," arXiv.org, Cornell University, April 29, 2012.

17. It should be noted that David Ricardo's famous comparative advantage model has long been discredited by many economists for its overly simplistic assumptions, a main fault being that the model assumes that offshoring would never occur. Yet Ricardo is still regarded as a saint in economic circles, and his trade theories are still taught as if they were canon law in economics departments at universities everywhere. I was taught to regard Ricardo's trade theories as sacrosanct and ironclad in my university Econ 101 lessons, and the press regularly repeats them as if they were obvious facts in opinion pages defending trade liberalization. But this is largely only true in the West; East Asian policymakers and scholars have long ignored Ricardo with great success.

18. Yakovenko, "Applications of Statistical Mechanics."

19. Anand Banerjee and Victor M. Yakovenko, "Universal Patterns of Inequality," *New Journal of Physics* 12 (July 2010).

20. Christine Lagarde, "Boosting Growth and Adjusting to Change" (speech given at Northwestern University, September 28, 2016, transcript by International Monetary Fund).

4. THE ANT FARM

1. Ginny Parker, "In Japan, Extinction Threatens an Ancient Empire of Industrious Insects," Associated Press, August 22, 1999.

2. Greg Yarrow, "The Basics of Population Dynamics," Clemson Extension, Fact Sheet No. 29, May 2009.

3. Joshua B. Fisher et al., "An Analysis of Spatial Clustering and Implications for Wildlife Management: A Burrowing Owl Example," *Environmental Management* 39, no. 3 (2007).

4. Diana Pilson and Mark D. Rausher, "Clumped Distribution Patterns in Goldenrod Aphids: Genetic and Ecological Mechanisms," *Ecological Entomology* 20, no. 1 (1995).

5. Deborah M. Gordon, *Ant Encounters: Interaction Networks and Colony Behavior* (Princeton, NJ: Princeton University Press, 2010).

6. Gordon, *Ant Encounters*.

7. Tool use in the animal kingdom is probably a lot more common than we realize. We already know that chimps use tools, as do otters and some species of birds. Tool use has even been witnessed in dolphins.

8. Erik Thomas Frank et al., "Saving the Injured: Rescue Behavior in the Termite-Hunting Ant *Megaponera analis*," *ScienceAdvances* 3, no. 4 (April 12, 2017).

9. Anna Dornhaus and Nigel R. Franks, "Individual and Collective Cognition in Ants and Other Insects," *Myrmecological News* 11 (2008).

10. Dornhaus and Franks, "Individual and Collective Cognition in Ants."

11. Gordon, *Ant Encounters*, 24.

12. Gordon, *Ant Encounters*, 24.

13. Gordon, *Ant Encounters*, 23.

14. However, China as a whole will benefit greatly.

15. Gordon, *Ant Encounters*, 84.

16. Shauna L. Price et al., "Recent Findings in Fungus-Growing Ants: Evolution, Ecology, and Behavior of a Complex Microbial Symbiosis," University of Kansas Department of Ecology and Evolutionary Biology and Smithsonian Tropical Research Institute, 2007.

17. Price et al., "Recent Findings in Fungus-Growing Ants."

18. Price et al., "Recent Findings in Fungus-Growing Ants."

19. Saori Watanabe, Jin Yoshimura, and Eisuke Hasegawa, "Ants Improve the Reproduction of Inferior Morphs to Maintain a Polymorphism in Symbiont Aphids," *Scientific Reports* 8, no. 2313 (February 2, 2018).

20. Tatiana Novgorodova, "Organization of Honeydew Collection by Foragers of Different Species of Ants (Hymenoptera: Formicidae): Effect of Colony Size and Species Specificity," *European Journal of Entomology* 112, no. 4 (2015).

21. Novgorodova, "Organization of Honeydew Collection."

22. Laura Elise Barbani, "Foraging Activity and Food Preferences of the Odorous House Ant," Virginia Polytechnic Institute and State University, June 2003.

23. Bernhard Stadler and Anthony F. G. Dixon, "Ecology and Evolution of Aphid-Ant Interactions," *Annual Review of Ecology, Evolution, and Systematics* 36 (December 2005).

24. Deborah M. Gordon, "The Queen Does Not Rule," *Aeon*, December 19, 2016.

5. COLONY DISORDER

1. Matthew Boesler and Yuko Takeo, "Why Japanification and Secular Stagnation Are Bad, Bad News," *Bloomberg Businessweek*, January 17, 2020.

2. Boesler and Takeo, "Why Japanification and Secular Stagnation."

3. Ross Douthat, "The Chinese Population Crisis," *New York Times*, January 18, 2020.

4. I am perfectly aware that they won't take my advice, of course.

5. Michal Krzyzanowski et al., "Air Pollution in the Mega-cities," *Current Environmental Health Reports* 1 (2014).

6. Bart Sweerts et al., "Estimation of Losses in Solar Energy Production from Air Pollution in China since 1960 Using Surface Radiation Data," *Nature Energy* 4 (July 8, 2019).

7. The World Bank, "Solid Waste Management," brief, September 23, 2019.

8. Moreno Di Marco et al., "Changes in Human Footprint Drive Changes in Species Extinction Risk," *Nature Communications* 9, no. 4621 (November 5, 2018).

9. Intergovernmental Science-Policy Platform on Biodiversity and Ecosystem Services, "Global Assessment Report on Biodiversity and Ecosystem Services," IPBES secretariat, Bonn, Germany, May 6, 2019.

10. Nancy B. Grimm et al., "Global Change and the Ecology of Cities," *Science* 319, no. 5864 (February 8, 2008).

11. J. L. Deneubourg, J. M. Pasteels, and J. C. Verhaeghe, "Probabilistic Behavior in Ants: A Strategy of Errors?," *Journal of Theoretical Biology* 105, no. 2 (1983).

12. Deneubourg, Pasteels, and Verhaeghe, "Probabilistic Behavior in Ants."

13. Janina Dobrzanska and Jan Dobrzanski, "The Foraging Behavior of the Ant *Myrmica laevinodis nyl.*," *Acta Neurobiologiae Experimentalis* 36, no. 5 (1976).

14. Dobrzanska and Dobrzanski, "The Foraging Behavior."

15. "More Children Killed by Unsafe Water, Than Bullets, Says UNICEF Chief," UN News, March 21, 2019.

16. Aimin Chen et al., "Developmental Neurotoxicants in e-Waste: An Emerging Health Concern," *Environmental Health Perspectives* 119, no. 4 (April 2011).

17. Alejandro G. Farji-Brener and Victoria Werenkraut, "The Effects of Ant Nests on Soil Fertility and Plant Performance: A Meta-analysis," *Journal of Animal Ecology* 86, no. 4 (July 2017).

18. A. N. M. Bot et al., "Waste Management in Leaf-Cutting Ants," *Ethology, Ecology & Evolution* 13 (2001).

19. Bot et al., "Waste Management."

20. M. Autuori, "Contribution to the Knowledge of the Sauva: The Sauveiro after First Flight," *Archives of the Institute of Biology* (Brazil) 18 (1947).

21. Adam G. Hart and Francis L. W. Ratnieks, "Waste Management in the Leaf-Cutting Ant *Atta colombica*," *Behavioral Ecology* 13, no. 2 (May 2001).

22. Hart and Ratnieks, "Waste Management."

23. Bot et al., "Waste Management."

24. Bot et al., "Waste Management."

25. Here "recycling" refers to an animal's directly collecting prior disposed-of trash or waste and reusing it somehow for its own benefit. Of course nature "recycles" waste all the time via bacteria and microorganisms that break waste down, allowing it to renter the ecosystem or food cycle, but that's an entirely separate process.

26. Alejandro J. Farji-Brener and Mariana Tadey, "Trash to Treasure: Leaf-Cutting Ants Repair Nest-Mound Damage by Recycling Refuse Dump Materials," *Behavioral Ecology* 23, no. 6 (2012).

27. Farji-Brener and Tadey, "Trash to Treasure."

28. Arindam Dhar, "Landfill Mining: A Comprehensive Literature Review," University of Texas at Arlington Technical Report, October 2015.

29. Dhar, "Landfill Mining."

30. Dhar, "Landfill Mining."

31. Joakim Krook et al., "Landfill Mining: A Review of Three Decades of Research," in proceedings of *The 14th European Roundtable on Sustainable Production and Consumption*, October 2010. Paper submitted by the Department of Management and Engineering, Environmental Management and Technology, Linköping University, Sweden.

32. The wind-energy boom in northwestern Texas and its revitalization of communities there demonstrates one possible exception.

CONCLUSION

1. Kate E. Langwig et al., "Context-Dependent Conservation Responses to Emerging Wildlife Diseases," *Frontiers in Ecology and the Environment* 13, no. 4 (May 1, 2015).

2. Chris Giles, "Economists among 'Least Trusted Professionals' in UK," *Financial Times*, November 3, 2019.

3. Mark Twain, *Life on the Mississippi* (Boston: James R. Osgood and Company, 1883).

4. The famed economist and frequent *New York Times* columnist Paul Krugman comes to mind here.

5. Charles Jones, "The End of Economic Growth? Unintended Consequences of a Declining Population," Stanford University Graduate School of Business, January 6, 2020.

6. Hiroaki Sasaki, "Non-renewable Resources and the Possibility of Sustainable Economic Development in a Positive or Negative Population Growth Econ-

omy," Munich Personal RePEc Archive, University Library of Munich, Germany, February 16, 2019.

INDEX

ABOUT THE AUTHOR

Nathanial Gronewold is a veteran international journalist with experience running assignments in Pakistan, Colombia, Kenya, Haiti, Singapore, South Korea, Taiwan, and beyond. He is the winner of the 2019 National Press Club Award for newsletter writing. Gronewold formerly reported on the United Nations and global affairs for *Nikkei, The Economist,* and The Canadian Press. He writes on environmental and energy news, establishing bureaus for E&E News in New York, Houston, and Japan. He has written close to three thousand articles, including those appearing in *Scientific American, Science Magazine,* and the *New York Times.* He is a two-time recipient of the Gold Prize for coverage of climate change from the United Nations Correspondents Association and received a 2012 honorable mention from the National Press Club. Both a professional journalist and an academic, Gronewold is pursuing his PhD in environmental science at Hokkaido University in Japan. He lives in northern Japan with his wife.